D. VAN NOSTRAND COMPANY, INC.
120 Alexander St., Princeton, New Jersey
(*Principal Office*)
24 West 40 Street, New York 18, New York

D. VAN NOSTRAND COMPANY, LTD.
358, Kensington High Street, London, W.14, England

D. VAN NOSTRAND COMPANY (Canada), LTD.
25 Hollinger Road, Toronto 16, Canada

IN THE UNITED STATES OF AMERICA

WILLIAM A. SHURCLIFF

Cambridge Electron Accelerator
Cambridge, Massachusetts

and

STANLEY S. BALLARD

University of Florida

POLARIZED LIGHT

Published for
The Commission on College Physics

D. VAN NOSTR

Princeton,

Toronto   *London*

Preface

The shock wave that is now spreading throughout the field of optics in general, and polarized light in particular, is the result not so much of an explosion as of an *implosion*. The laser, our most intense laboratory source of polarized light, was invented by researchers in electronics and microwaves. Botanists have discovered that the direction of growth of certain plants can be determined by controlling the polarization form of the illumination, and zoologists have found that bees, ants, and various other creatures routinely use the polarization of sky light as a navigational "compass." High-energy physicists have found that the most modern particle accelerator, the synchrotron, is a superb source of polarized x-rays. Astronomers find that the polarization of radio waves from planets and from stars offers important clues as to the dynamics of those bodies. Chemists and mechanical engineers are finding new uses for polarized light as an analytical tool. Theoreticians have discovered short-cut methods of dealing with polarized light algebraically. From all sides the inrush of new ideas is imparting new vigor to this classical subject.

The present book explains briefly the foundations of polarized light, indicates the power of the short-cut theoretical methods due to H. Poincaré, G. G. Stokes, H. Mueller, and R. C. Jones, surveys the time-honored applications of polarized light, and then describes some of the newest and most dramatic applications.

We say little about the algebraic descriptions of waves since this subject is covered so clearly in R. A. Waldron's *Waves and Oscillations* (Momentum Book #4). Likewise we say little about the applications of polarized light to crystallography since

3

E. A. Wood's *Crystals and Light* (Momentum Book #5) explains the subject so fully and provides such a fine set of color photographs of the interference patterns observed.

The present authors, in their years of association with the Polaroid Corporation (led by Dr. Edwin H. Land, the inventor of the first high-quality, low-cost polarizers) and in their associations with various university and government laboratories, have been fortunate witnesses of many of the new developments. Grateful thanks are due countless persons, within and outside the field of optics, who have pointed out to us and explained to us some of the more dramatic advances. We are much indebted to Drs. C. D. West and R. Clark Jones of Polaroid for help and instruction given over a period of many years. We wish to thank the Harvard University Press for permission to use several of the illustrations from the book *Polarized Light: Production and Use* by W. A. Shurcliff, published in 1962.

Cambridge, Massachusetts   WILLIAM A. SHURCLIFF
Gainesville, Florida   STANLEY S. BALLARD

Table of Contents

Plates (following p. 32)

1 *Polarization: A Simplifying Process*

POLARIZATION AND SIMPLICITY

Polarized light isn't really polar. It is merely simple.

Various kinds of simplicity are already familiar to students of optics—for example, simplicity with respect to wavelength. A beam of white light is far from simple since a great variety of wavelengths are included; but a prism can disperse such a beam, spreading it in a "fan" of directions each of which corresponds to a single wavelength. By dealing with monochromatic light, physicists make much progress: they can discover the gross structure of atoms, or detect rare elements in the sun.

The simplification can be carried one stage further: one can subdivide a given beam until it consists of just one wavelength and *just one polarization form*. With such a beam—the end of the road in simplicity—one can perform hundreds of experiments of even more delicate and revealing types. One can search deeper into the structure of the atom by distinguishing differences in polarization of certain spectral lines. One can determine the pattern of magnetic fields on the sun by noting the pattern of polarization in the emitted light.

More important, *polarized* monochromatic light, being the very simplest kind of light, gives man his closest look at what, ultimately, light really is.

WHAT IS LIGHT?

The bald fact is that no one knows what light really consists of. Textbooks on light give no answer—instead they hedge. The more advanced the book, the more carefully the author avoids saying what light *is*.

However, much is known as to how light behaves. Two main patterns of behavior are found:

(1) In many respects it behaves as if it were a train of waves;

(2) In other respects it behaves as if it were a random group of tiny particles (photons).

Using the wave picture and the mathematics of waves, one can solve many kinds of problems accurately; but in other problems one arrives at answers that are utterly false. Accurate solutions are found to problems that involve intense beams of light—for example, a beam of sunlight that strikes a lens or a prism or a set of slits; using wave theory one can calculate rather accurately what fraction of the light will go in any given direction.

But if the beam is feeble and of short duration, the wave theory fails. It predicts a distribution that is just as smooth as when the beam is intense; yet in fact the distribution appears granular. Indeed, if the light is extremely weak, it acts as if it were concentrated in just a few widely separated points. Also, wave theory fails in calculations as to how much energy the beam imparts to a given atom that it strikes. Wave theory suggests that a more intense beam will impart more energy to the atom. But experiment shows this to be false: the more intense beam may affect a larger number of atoms, but the vigor of the effect remains unchanged.

Using the particle (photon) model, a physicist succeeds brilliantly in some problems and fails in others. He succeeds when attempting to predict the amount of energy delivered in any one elementary act such as the absorption or scattering of light by an individual atom; but he cannot predict *where* the absorption will occur.

Accordingly, the wave theory is favored by persons who deal with large quantities of photons (e.g., persons working with intense visible light, or infrared radiation, or radio signals), and the photon theory is preferred by persons who deal with small numbers of very energetic photons (e.g., persons working with gamma rays from a bit of radioactive material).

Waves or paticles? The controversy has gone on for half a century, and the present authors intend to keep out of it. Nearly any positive statement one makes is likely to be wrong. But this much is certain: respectful attention is merited both by the wave-like characteristics and by the particle-like characteristics. Both are useful models—brilliant models. But each is dead wrong in certain situations. Happily, the two models complement one another.

TRANSVERSE NATURE OF THE WAVES

A striking fact about light waves is that they are transverse. The displacements, of electromagnetic nature as explained in Chapter 2, are not along the line of travel, but perpendicular to it. For example, if the direction of travel of a given light beam is east, the electric vibrations may be up and down, or north and south, or along some other line perpendicular to the east-west axis. Chapter 2 deals with this subject.

DEFINITION OF POLARIZED LIGHT

Polarized light is light whose transverse vibration has a simple pattern, such as that exemplified in Fig. 1-1a. Here the light is

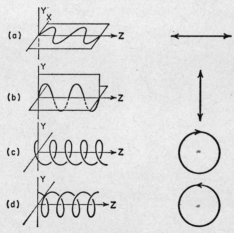

FIG. 1-1 Snapshot pattern (left column) and sectional pattern (right column) of a beam that is polarized (a) linearly, horizontally, (b) linearly, vertically, (c) right circularly, (d) left circularly. In viewing a sectional pattern, the observer is assumed to be situated far to the right, on the positive Z axis, looking toward the light source at the origin.

traveling east along the Z-axis of a right-handed coordinate system. The electric vibrations are horizontal, and accordingly the light is said to be polarized linearly and horizontally. The left portion of Fig. 1-1a shows the *snapshot pattern* of the wave train; that is, it shows how the train might appear in a snapshop photograph taken with an extremely fast-acting camera. The right portion shows the stylized *sectional pattern,* which corresponds to a time-exposure photograph taken by a camera looking at a cross section of the beam, i.e., situated on the axis of the beam and aimed straight at the light source. Of course, both types of patterns are in a sense fictitious, since there is no way of actually seeing or photographing the waves as such. But the patterns are invaluable as models.

When the transverse vibrations are vertical (Fig. 1-1b) the sectional pattern is indicated by a short vertical line and the light is said to be polarized linearly and vertically.

For some beams the snapshot pattern is a helix and the sectional pattern is a circle; such a beam is said to be circularly polarized. There are two possibilities here: the helix may be of right-handed type (like the thread of an ordinary machine screw) or of left-handed type. As suggested by Figs. 1-1c and 1-1d, these two types are basically different; they appear different irrespective of the viewpoint of the observer. Light having a right-handed helical pattern is called right circularly polarized light; more conveniently it is represented as a circle traced in clockwise sense. To see that the two representations are equivalent one may note that the intersection point of a right helix and a fixed plane describes a clockwise circle as the helix is pushed forward (without rotating) through the plane and toward the observer.

Some beams have a snapshot pattern resembling a transversely flattened helix and a sectional pattern that is elliptical. Such beams are called elliptically polarized. They too may be right-handed or left-handed.

Figure 1-2 presents a wide variety of sectional patterns.

Different definitions of polarized light are used in the photon theory, as is indicated in Chapter 8. For example, one may employ a set of four numbers called the components of the Stokes vector.

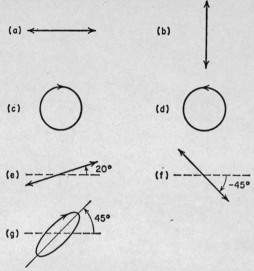

FIG. 1-2 Sectional pattern of a beam polarized (a) linearly, horizontally, (b) linearly, vertically, (c) right circularly, (d) left circularly, (e) linearly at 20°, (f) linearly at −45°, (g) right elliptically at 45°.

DIFFERENCES BETWEEN LIGHT AND SOUND

The temptation to regard light and sound as nearly alike must be resisted. There are resemblances between them, and in some kinds of problems similar diagrams and similar equations are used. But the differences are impressive. The propagation of sound waves in air is a gross, statistical process, without scientific implications as to the nature of matter or energy. The propagation of light is an "ultimate," which reaches to the very heart of physics, cosmology, and philosophy—so much so, in fact, that we shall probably never have an adequate understanding of light until we understand those other giant enigmas: matter, energy, and vacuum.

Sound cannot travel in a vacuum. Light travels best in a vacuum.

Sound waves in air are longitudinal waves; they have no sectional pattern at all, and hence are never polarized. Light waves are transverse and may display a variety of linear, circular, or elliptical sectional patterns.

UNPOLARIZED LIGHT

Unpolarized light—i.e., nonpolarized light—is a grand mixture, a mess. To define it is difficult; to depict it in a diagram is impossible. The best definition is the following, couched in negative terms: Unpolarized light is light that exhibits no long-term preference as to vibration pattern (sectional pattern). In other words, when one can find no preponderance of any one vibration direction over any other vibration direction, and when one can find no preponderance of clockwise vibration over counterclockwise vibration or vice versa, one calls the light unpolarized.

Some authors attempt to depict unpolarized light by means of a diagram such as is shown in Fig. 1-3. They draw radial lines

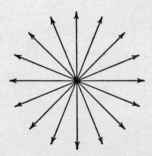

FIG. 1-3 Unsuccessful attempt to depict unpolarized light. The diagram is inadequate in several ways, such as failing to include circles and ellipses.

in many directions, to indicate that no one direction predominates. But they should include also some circles and ellipses, and indicate an equal abundance of clockwise and counterclockwise forms. To be completely accurate they should present a series of sketches, each different from the others, to show that the pattern is constantly changing from moment to moment. At any one moment any beam is likely to have some dominating pattern of vibration, if only by accident; but the pattern of an unpolarized beam can be thought of as continually changing, favoring now one pattern, now another, so that over a long period of time no over-all preference is manifest.

PARTIALLY POLARIZED LIGHT

The polarization of light is not an all-or-none affair. Intermediate degrees of polarization may exist. When one mixes a 1-watt beam of polarized light with a 1-watt beam of unpolarized light, the resulting beam is called partially polarized, and the degree of polarization is said to be $\frac{1}{2}$, or 50 percent. If one mixes I_p watts of polarized light with I_u watts of unpolarized light, the degree of polarization is $I_p/(I_p + I_u)$.

This is merely a matter of definition, i.e., a conventional way of describing the intermediate extent of polarization. A clear convention is needed since nearly all actual beams are partially polarized, as is pointed out in Chapter 10.

ANALYSIS OF PARTIALLY POLARIZED LIGHT

There are two approaches to analyzing partially polarized light into component parts: a less interesting approach and a more interesting approach. The former is easy. The latter is impossible to accomplish physically, but provides food for thought.

Choosing the less interesting approach, called the vibration-form dichotomy, one divides the beam into two portions, each of which is completely polarized; the two are always polarized oppositely (orthogonally) to each other. The Wollaston prism, discussed in the next chapter, makes such a division. It has the capability of dividing a partially polarized beam (indeed, *any* beam) into a *horizontally* linearly polarized portion and also a *vertically* linearly polarized portion. Or, if the prism is turned to some other angle, such as 15°, it produces two portions linearly polarized at 15° and 105°. No matter what its orientation, it produces two beams that are polarized linearly and orthogonally. Thus it is called a linearly polarizing beam-splitter. The two emerging beams have slightly different directions of travel, and can be used separately; hence the importance of this famous device. Having practically no absorption for visible light, it makes the analysis with almost 100 percent efficiency.

Circular beam-splitters also exist. They divide beams into two circularly polarized portions having opposite sense (opposite handedness). Circular beam-splitters are expensive, and are sel-

dom used. (The word *circular* here denotes, of course, the shape
of the vibration pattern of an emerging beam. It does not imply
that the device has been machined so as to have the shape of a
disk.)

Choosing the more interesting approach, one divides the
partially polarized beam into a completely polarized portion
and a completely unpolarized portion. That is, one makes a
polarized-vs-unpolarized dichotomy. Usually, the unpolarized
portion is not wanted; one would like to separate it and discard
it. Unfortunately, however, there is no known way of doing this.
Many physical devices can distinguish different forms of polarization, but none can separate polarized light from unpolarized
light.

JEKYLL AND HYDE NATURE OF POLARIZED LIGHT

There is perhaps nothing that can be transformed so drastically, quickly, and efficiently as polarized light. When a beam of
circularly polarized light strikes a Wollaston prism, the two
emerging beams are found to be *linearly* polarized. The change
from circular to linear form occurs in less than the time required
for the light to pass through the prism, i.e., less than 10^{-10}
second. The efficiency is practically 100 percent.

Actually, any form of polarized light can be converted to any
other form with this same speed and efficiency.

The quick change in polarization form is especially impressive
in terms of the photon theory of light. Apparently photons of
one elementary type strike the beam-splitter, and photons of a
different elementary type emerge. How can this be? How can
something elemental be forced into a different mold? No satisfactory explanation (involving photons) is known. One can
merely say that the beam-splitter compels the incoming photons
to adopt a new guise; each incoming circularly polarized photon,
say, is forced to become either a horizontally linearly polarized
photon or a vertically linearly polarized photon, just as a British
Labour-party man visiting the United States might be asked to
register as a Democrat or a Republican. Such sudden, forced
choices are the rule, not the exception, in quantum mechanics,

and no finer illustration can be found than in the instantaneous conversion of polarized light.

PHOTONS: UNIQUE AMONG FUNDAMENTAL PARTICLES

We close this introductory chapter with some reminders of the unique properties of light in general and of polarized light in particular.

Thirty-five years ago, only three kinds of fundamental particles were known: electrons, protons, and photons. The photons were of outstanding interest.

Today there are more than 35 known kinds of particles—the original three and also neutrinos, muons, pi mesons, kaons, lambdas, sigmas, and several others. Yet photons continue to be of outstanding interest.

What is so special about photons? A hundred pages could not provide a full answer; a brief answer is as follows:

Visibility: Photons are the only particles that man is "directly wired" for. Thanks to our eyes, we can *see* photons. Furthermore, we can judge them with great accuracy. We can distinguish differences in direction-of-travel with an accuracy of a hundredth of a degree; we do this whenever we see the separate pickets in a fence a few blocks away. We can distinguish wavelength differences of 1 percent; we do this when we distinguish a reddish necktie from a reddish-orange necktie. We can perceive a great variety of directions and wavelengths simultaneously, as when we watch a three-ring circus. We can cope with light so intense that 10^{15} photons enter the eye in a second; yet we can still see fairly well when the light is so feeble that only a few thousand photons enter the eye in a second—from a distant star, say. As few as a dozen photons can be seen when the observer is dark-adapted and all the photons reach the same part of the retina at the same time. (Few other detectors known have such extreme sensitivity.) Man has *two* of these photon-detectors (two eyes), hence can receive stereoscopic impressions; that is, he can "see depth." The detectors require no conscious effort at focusing or servicing, and often perform essentially perfectly for a lifetime.

To detect other elementary particles is far more difficult. Special instruments are required, some of which are very costly. A modern bubble-chamber for detecting and analyzing pions may cost $1 million. What would a high-energy physicist not give to be able to make these analyses instantly, by eye!

Ease of Production: Photons are easy to produce. When you light a match, you produce trillions of them. Even a firefly produces billions. A 20¢ light bulb will emit an enormous stream of photons for months. The stream can be quickly switched on or off. Yet to produce lambda particles, say, an experimenter needs millions of dollars' worth of equipment.

Wide Range of Energy: Photons of any desired energy from almost zero electron volts (0 eV) to many billions of electron volts can be produced. A photon of green light has an energy of 2.3 eV; infrared photons have slightly less energy, and the commonest radio signals consist of photons having billion-fold smaller energies. High-energy synchrotrons can produce particles that, on colliding, emit photons that are billions of times *more* energetic than green-light photons.

But in producing pions, for example, one usually achieves a smaller range of energies. There is a threshold energy ("rest-mass energy") of many millions of electron volts to be surmounted, and accordingly the reliable production of beams of very-low-energy pions is difficult.

Ease of Manipulation: Photons can be reflected by a piece of aluminum foil, refracted by wedge-shaped pieces of broken glass or by a piece of cracked ice. Different energies can be selected with the aid of bits of colored cellophane. Laboratory supply-houses provide, for $50 or less, an assortment of devices that serve for dozens of such experiments.

For muons, pions, protons, etc., there is no practical mirror or prism. Merely to stop the particles may require thick blocks of lead. To sort out the different energies requires use of electro-magnets weighing many tons. Every experiment is a long and costly undertaking. (See David H. Frisch and Alan M. Thorndike, *Elementary Particles,* Momentum Book #1, D. Van Nostrand Company, Inc., Princeton, N. J., 1964).

Safety: Light beams are harmless, ordinarily. When you light a match, you need no lead shielding, no license from the Atomic

Energy Commission. The light from powerful lasers, however, may be so intense as to blind a person instantly.

Laboratories producing pions, muons, etc., are places of danger. Hundreds of tons of concrete shielding blocks must be employed, together with many other safety measures.

Creativity: The creative ability of photons almost staggers the imagination. Recent work with high-energy accelerators suggests that photons of sufficiently high energy can be used to produce—directly or indirectly—all other known types of particles. For example, the 6 billion-electron-volt photons produced by the Cambridge Electron Accelerator at Cambridge, Mass., can create muons, pions, kaons, protons, neutrons, lambdas, and sigmas. The several-times-more-powerful accelerator now being built at Stanford University in California is expected to provide photons that can create any type of fundamental particle so far detected by man.

For all these reasons, photons continue to fascinate the scientist. Whether working with radio telescopes, infrared detectors, visible light, x-rays, or strange types of antiparticles, he continues to marvel at this simplest and most versatile of the fundamental particles.

2 *Assistance Provided by the Electromagnetic Theory*

BASIC CONCEPTS

We shall mention only those parts of the electromagnetic theory of light that help one understand the basic features of polarized light and assist in solving practical problems.

According to the electromagnetic theory, light consists of waves of the *same type* that are produced when electric charges are moved back and forth rapidly. When one grasps an electro-statically charged glass rod and shakes it back and forth, the electric field a foot or two away fluctuates. If one could shake the rod millions of times a second, the fluctuations could be detected many miles away. Rapidly oscillating magnets, also, can produce disturbances that travel to great distances.

Maxwell's suggestion, in 1873, that light belongs to this same family of disturbances startled the scientific world. The first strong evidence in support of his idea was the experimental demonstration that the speed of light is the same as that of dis-turbances produced by oscillating charges. Countless additional experiments have provided further confirmation.

Maxwell's discovery paved the way for rapid advances in man's ability to predict the behavior of light. The same concepts and equations that had been developed for electromagnetic disturb-ances could be applied immediately to light.

What are the basic concepts of the electromagnetic theory? Some of the principal ones are stated below in nonmathematical form:

(a) The disturbances consist of trains of waves.
(b) Electrical waves are always accompanied by magnetic waves, and vice versa.
(c) When traveling in empty space, the waves always travel at the same speed, namely 3×10^{10} cm/sec, usually designated by c.
(d) The waves are transverse; that is, the displacements are per-pendicular to the direction in which the wave-train is traveling.
(e) At any one point in empty space the electric and magnetic dis-placements E and H are mutually perpendicular.
(f) As the waves spread outward into empty space, their amplitudes decrease steadily.
(g) The waves carry energy, and at any given location the rate of flow of energy (the power) is proportional to the square of the amplitude, e.g., of the electric wave.
(h) When two disturbances are superimposed, they combine in the simplest possible way: at any given location and given instant, the combined displacement is the vector sum of the two indi-vidual displacements.
(i) Most important: light, too, conforms to all these rules. It is a regular member of the electromagnetic-wave family, and differs from radio waves, for example, only in that light has a much shorter wavelength and a correspondingly higher frequency.

How can a student of polarized light make practical use of these concepts? He can do so in (1) describing a disturbance, (2) computing the effect of combining two disturbances, and (3) computing the effect of subdividing a disturbance. These three processes are central to the understanding of polarized light. Let us, therefore, consider each one separately.

DESCRIBING A DISTURBANCE

To describe a disturbance (i.e., a light beam) in terms of the electromagnetic theory is easier than one might suppose. Indeed, at the outset one usually makes the following simplification: one describes *just the electric waves*. In most problems, a solution that is valid for the electric waves is valid also for the magnetic waves; hence to describe the magnetic waves would be redundant. One could choose to describe either the electric or the magnetic waves, but certain experiments in photochemistry indicate that it is the electric part of the combined waves that interacts directly with matter. The *theoretical* importance is the same; the electric and magnetic waves are, so to speak, different sides of the same coin.

A second commonly made simplification is the assumption that the train of waves is monochromatic, i.e., involves but a single wavelength λ.

A third simplification is the assumption that one is concerned with but a small region of space, and that this region is so far from the light source that the amplitude throughout the region is practically constant.

Accordingly, one can depict the wave train—at a given instant —by a sine curve such as that shown in Fig. 2-1. A cosine curve

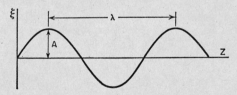

FIG. 2-1 Simplest representation of a monochromatic train of waves at one particular instant. A is the amplitude and λ is the wavelength.

is equivalent, of course, except that a different starting point (different epoch angle, or phase) is implied. If one regards the displacement as being a true geometrical displacement, one must regard the magnitude, or scale, of the amplitude A in the diagram as being arbitrary, since no information exists as to the absolute size of the displacement. When one regards the displacement as implying a certain strength and direction of electric field, one can express the magnitude of A in terms of volts per centimeter.

If one prefers analytical geometry to pictures, he can specify the disturbance thus:

$$\xi = A \sin \left(\omega t - \frac{2\pi Z}{\lambda} \right)$$

Here A is the amplitude, ω is the angular frequency (that is, 2π times the frequency v in cycles per unit time), and λ, called the wavelength, is the distance from one crest to the next. The entire expression, represented by ξ, is the displacement (or, more specifically, the magnitude of electric field strength) at any given time t and any given position Z along the axis of propagation.

Persons familiar with complex numbers often prefer to use an equivalent expression such as

$$\xi = A e^{i \left(\omega t - \frac{2\pi Z}{\lambda} \right)}$$

or, more briefly,

$$\xi = A e^{i\phi}$$

where ϕ stands for the entire phase angle $\left(\omega t - \frac{2\pi Z}{\lambda} \right)$. The displacement is identified arbitrarily with the *real part* of ξ.

To indicate that the displacement is parallel to, say, the X-axis (rather than the Y-axis, or some other direction perpendicular to the propagation direction), one may simply insert the statement "displacement is parallel to the X-axis." Or one may write the symbol **i** in front of the algebraic expression and state that **i** is a unit vector pointing in the direction of the positive axis is X. To indicate a displacement parallel to the Y-axis, one would include the symbol **j**.

Circularly polarized beams and elliptically polarized beams can also be described algebraically, by methods indicated in a later paragraph.

COMPUTING THE EFFECT OF COMBINING TWO DISTURBANCES

Suppose that we combine two beams of light that have the same wavelength and same phase, and suppose that the two beams are traveling along the same path (Z-axis) and have identical directions of displacement (direction parallel to the X-axis). Suppose, finally, that the two amplitudes, called A_1 and A_2, differ. According to electromagnetic theory, when the two beams are combined, the resulting disturbance ξ_c corresponds to the simple addition of the two initial disturbances. That is:

$$\xi_c = A_1 \sin\left(\omega t - \frac{2\pi Z}{\lambda}\right) + A_2 \sin\left(\omega t - \frac{2\pi Z}{\lambda}\right)$$

$$= (A_1 + A_2) \sin\left(\omega t - \frac{2\pi Z}{\lambda}\right)$$

$$= A_c \sin\left(\omega t - \frac{2\pi Z}{\lambda}\right), \text{ where } A_c \text{ stands for } (A_1 + A_2).$$

Or, using the complex expressions,

$$\xi_c = A_1 e^{i\phi} + A_2 e^{i\phi} = (A_1 + A_2)e^{i\phi} = A_c e^{i\phi}.$$

In words: The resulting disturbance is the same as either initial disturbance except that the amplitude is greater, being the sum of the two initial amplitudes.

The slightly more complicated case in which the second initial disturbance has a 90-degree greater epoch angle, or phase, than the first initial disturbance is easily discussed. Here the sum of the two disturbances is indicated by an expression such as

$$A_1 e^{i\phi} + A_2 e^{i\left(\phi + \frac{\pi}{2}\right)}.$$

If the phase of the second disturbance relative to the first has some general value γ, the sum of the disturbances is:

$$A_1 e^{i\phi} + A_2 e^{i(\phi+\gamma)}.$$

If γ happens to be 180°, or π radians, the second term in the above expression, in this case $A_2 e^{i(\phi+\pi)}$, can be simplified with interesting consequences. It can be written $(A_2e^{i\phi})(e^{i\pi})$; and remembering the identity $e^{i\pi} = -1$, one can reduce the expression to $-A_2e^{i\phi}$. Then the entire expression becomes

$$A_1e^{i\phi} - A_2e^{i\phi} \quad \text{or} \quad (A_1 - A_2)e^{i\phi}.$$

This means that the resulting disturbance is the same as either initial disturbance except that the resulting amplitude is the *difference* between the two initial amplitudes A_1 and A_2. (The same result is reached equally easily using the sine notation; but in more complicated problems the exponential notation proves to be simpler.) In particular, if A_1 and A_2 happen to be equal, the difference is 0, and the resulting disturbance vanishes entirely! This outcome is frequently observed in experiments designed to demonstrate the interference of polarized light.

Suppose, now, the two initial disturbances have the same phase but have displacements that are parallel to the X- and Y-axes respectively. Then the expression for the sum might be indicated with the aid of unit vectors, thus:

$$(\mathbf{i}A_1)(\text{real part of } e^{i\phi}) + (\mathbf{j}A_2)(\text{real part of } e^{i\phi})$$

or, more briefly,

$$(\mathbf{i}A_1 + \mathbf{j}A_2)(\text{real part of } e^{i\phi}).$$

By drawing the two vectors $(\mathbf{i}A_1)$ and $(\mathbf{j}A_2)$ and finding the resultant as in Fig. 2-2, one finds the direction of the displacement of the resulting beam and also the amplitude of this beam. (One must avoid confusing the symbol i, used in complex notation, with the unit vector $\mathbf{i}$. They have utterly different meanings.)

It is instructive to consider a similar situation, except that the phase of the vertically vibrating initial beam is 90° greater than that of the horizontally vibrating initial beam. Here the combined displacement may be written

$$\mathbf{i}A_1e^{i\phi_1} + \mathbf{j}A_2e^{i\left(\phi_1 + \frac{\pi}{2}\right)}$$

where, in the interest of brevity, the qualifying phrase "real part of" is omitted. The two components fluctuate in different phases; at a given location the displacement of one is a maximum when that of the other is zero, and vice versa. Thus the resultant vector

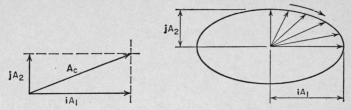

(Left) FIG. 2-2   Graphical method of finding the result of combining two beams that have the same wavelength and same phase, but have different amplitudes A_1 and A_2 and have mutually perpendicular vibration directions i and j. The resultant A_c indicates the direction and magnitude of the sectional pattern of the combined beam. (Right) FIG. 2-3 Method of finding the elliptical sectional pattern that results when the two initial beams have different amplitudes A_1 and A_2, vibrate in mutually perpendicular planes, and when the vertically vibrating initial beam has a 90° greater phase than the other initial beam. The succession of arrows represents resultants found at successive moments, and the tips of the arrows trace out an ellipse with clockwise sense.

has different directions at different moments, as suggested by Fig. 2-3. The tail of the vector is thought of as being fixed, and the head of the vector then occupies a succession of positions; it traces out an ellipse in *clockwise* manner as judged by an observer situated far out along the Z-axis and looking toward the light source at the origin of coordinates. Accordingly the combined beam is called right elliptically polarized.

If when the phase difference is 90° the two amplitudes A_1 and A_2 are assumed equal, the sectional pattern is a circle; the combined beam is then called circularly polarized. Conversely, any circularly polarized beam can be expressed algebraically as the sum of two linearly polarized beams that have equal amplitudes, vibrate in mutually perpendicular planes, and have phases that differ by 90°. The algebraic expression for circularly polarized light is:

$$\mathbf{i}Ae^{i\phi} + \mathbf{j}Ae^{i\left(\phi \pm \frac{\pi}{2}\right)}.$$

Employing the identity $e^{i\pi/2} = i$, one may reduce the expression to:

$$Ae^{i\phi}[\mathbf{i} + \mathbf{j}(\pm i)].$$

If the combined beam is right-handed, one employs the term $(+i)$; if it is left-handed, one employs $(-i)$.

If the phase difference lies between 0 and 90°, the sectional pattern is an ellipse that is tilted "upward to the right, downward to the left," as in Fig. 1-2g. If it lies between 0 and −90°, the tilt is downward to the right, upward to the left.

✗ In summary, adding any two linearly polarized beams that have a fixed phase relationship leads to a combined beam having a sectional pattern that is linear, circular, or elliptical, depending on the vibration directions, amplitudes, and phases involved. The result of the combining process can be found quickly and moderately accurately with the aid of a simple sketch, and may be found with complete accuracy by adding the appropriate algebraic expressions. When moderate accuracy suffices, the sketch is preferred. The validity of the graphical and algebraic methods is guaranteed by the electromagnetic theory and confirmed by experiment. The methods permit a twentieth-century student to solve, quickly and with confidence, problems that baffled scientists of 150 years ago.

✗ The foregoing discussion is valid only if the beams in question are *coherent*. Two beams are said to be coherent if, when superimposed, they produce an interference pattern. If the intensities found at the maxima and minima correspond to $(A_1 + A_2)^2$ and $(A_1 - A_2)^2$, where A_1 and A_2 are the amplitudes of the individual beams, the beams are said to be 100 percent coherent. If the beams are *incoherent*, no interference pattern is produced, and the combined intensity has the value $A_1{}^2 + A_2{}^2$; the squaring operation is to be performed before, rather than after, the two terms are combined. Two beams cannot be 100 percent coherent unless they have the same spectral energy distributions, have a fixed phase relationship at specified fiducial points, and have the same degree of polarization. If the polarization forms differ, a suitable retardation plate may be introduced in one beam to make the polarization forms alike before the interference pattern is evaluated. The result of combining beams that are incoherent is usually found with the aid of the Stokes vectors, discussed in Chapter 8.

COMPUTING THE EFFECT OF SUBDIVIDING
A DISTURBANCE

A person familiar with electromagnetic theory and with the rules governing the combining of polarized beams recognizes that any polarized beam can be *subdivided* into two beams that are linearly polarized along any desired pair of mutually perpendicular directions. Let us suppose the given beam has a sectional pattern and amplitude indicated by vector A_c in Fig. 2-2, and one wishes to subdivide it, or analyze it, into beams whose sectional patterns have the directions of the X- and Y-axes. To predict the outcome, one merely draws perpendicular lines from the tip of vector A_c to the two axes and finds the intercepts A_1 and A_2 thereon. The lengths of these intercepts indicate the amplitudes of the horizontally polarized and vertically polarized components. The *squares* of the intercepts, i.e., A_1^2 and A_2^2, indicate the respective energies (powers).

The outcome can be verified experimentally by passing the original beam through a Wollaston prism or other linearly polarizing beam-splitter. One orients the prism so that its crystallographic axes are aligned with the desired X- and Y-axes, as indicated in Fig. 2-4. Two beams emerge from the prism. One is

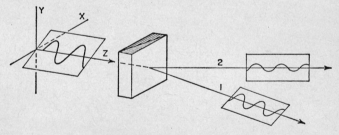

FIG. 2-4 Manner in which a Wollaston prism, oriented with respect to the *X*- and *Y*-axes, divides a linearly polarized beam vibrating in an oblique plane into two beams 1 and 2 having vibration directions parallel to the *X*- and *Y*-axes respectively.

polarized horizontally, the other vertically. The amplitudes and powers are in full accord with the predictions.

The variety of analyses that can be made is great. A linearly polarized beam can be analyzed—not merely into any desired orthogonal pair of *linear* polarization forms, but into clockwise and counterclockwise *circular* forms, or into orthogonal pairs of elliptical forms. Again, one usually employs graphical methods, but algebraic methods may be used if great precision is needed.

We are now finished with the elementary account of what polarized light is and how to predict the result of combining or subdividing beams. We are ready to consider the methods by which polarized light is produced.

3   *Dichroic Polarizers*

PEDESTRIAN NATURE OF POLARIZERS

Before launching into a discussion of polarizers, we must emphasize that the most elegant way of producing polarized light does not require any polarizer at all. One can produce light that is "born polarized," i.e., polarized from the very outset. Chapter 12 deals with such "polarized light generators," which range in size from atomic to galactic.

In the laboratory the usual method of producing polarized light is the pedestrian, three-step method of (1) producing ordinary unpolarized light, (2) dividing it (actually or at least in principle) into two orthogonally polarized components, and (3) eliminating one of these components. The component that remains is then available for use. The process is inefficient, but cheap and easy.

The device that divides the unpolarized light into two components and discards one is called a polarizer. (If it separates the two components and preserves both, it is a polarizing beam-splitter.) If the discarding process is imperfect, so that a fraction of the unwanted component survives, the polarizer is called a partial polarizer, or a leaky polarizer.

Clearly, a polarizer is really a *divider-and-selector*. It does not *create* transverse vibrations, but merely divides (resolves) the existing vibrations into two classes and selects one. Perhaps the most provocative fact is that it divides the incident beam into just two classes, never more.

Many types of polarizers exist. Why? Because so many types of optical manipulations may be used: absorption, reflection, refraction, or scattering. Any of these processes may be used to resolve a beam into polarized components. The key to the polarization process is asymmetry—e.g., structural asymmetry within the polarizer, or asymmetry (obliquity) of the mounting of the polarizer, or asymmetry of the viewing direction relative to the direction of the incident beam.

In this chapter we consider the most easily understood class of polarizer: that which employs asymmetry of absorption, or *dichroism.* Perhaps the simplest member of this family is the wire-grid-type polarizer.

THE WIRE-GRID POLARIZER

This device consists of an array of slender wires arranged parallel to one another as indicated in Fig. 3-1. Being metallic,

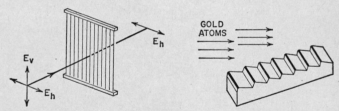

(Left) FIG. 3-1 How the wire-grid-type polarizer works. The incident, unpolarized beam contains vertical and horizontal components E_v and E_h of electric field. The former produce currents in the vertical wires and are absorbed. The latter produce no currents, hence pass through freely. (Right) FIG. 3-2 Evaporated gold atoms stream through a vacuum and strike the left sides of the ridges, or rulings, of the plastic diffraction grating. Thus thousands of parallel, submicroscopic "wires" of gold are produced in a few minutes.

the wires provide high conductivity for electric fields parallel to the wires. Such fields produce electric currents in the wires, and the energy of the fields is converted to energy of the currents. The latter is then converted to heat, because of the small but significant electrical resistance of the wires.

However, because of the nonconducting spaces between the wires, no current can flow perpendicular to them. Hence electric fields perpendicular to the wires produce no currents and lose no energy.

Thus the wire grid, when placed in an unpolarized beam, drains the energy out of one component and allows the other component to pass through with almost no decrease (no attenuation) at all. The transmitted component is the one whose electric vibration direction is perpendicular to the wires. (If the wires were arranged randomly in all directions, both components would be absorbed and the device would transmit no light at all.)

Although easy to understand, the wire-grid polarizer is difficult to make. Indeed, no one succeeded in making a really satisfactory wire-grid polarizer until 1963, when the writers' former colleagues G. R. Bird and M. Parrish amazed optical men by overcoming the almost insuperable difficulty of producing wires less than a wavelength of light in diameter and separated by uniform gaps less than a wavelength wide. No conventional methods of wire-making could suffice, and no micro-manipulator could arrange the wires so precisely, since 30,000 wires are needed in a polarizer scarcely larger than a postage stamp.

The problem was solved by evaporating metal onto the ridges of a transparent, 50,000-lines-per-inch diffraction grating. As indicated in Fig. 3-2, a stream of gold atoms in a large evacuated chamber is aimed almost at grazing incidence onto the grating in such a way that each ridge becomes coated along one side. (The valleys are in the shadow and receive no metal.) The coatings, only a few hundred atoms thick, are more nearly flat than round in cross section; they are too small to be seen even under a high-power optical microscope, but can be photographed with an electron microscope. Despite their small cross section, they provide good conductivity in one direction and complete insulation in the orthogonal direction.

A wire grid performs very badly if the wires are thick and the spaces between them are large. A thick wire permits some flow of current transversely across the individual wire, and thus even the wanted component of the light is absorbed to some extent. Wide spaces between wires permit some of the unwanted component to leak through. Wire grids for polarizing 3-cm microwaves are easy to make, since it is only necessary that the wire diameter and spacing be small compared to 3 cm. For visible light, however, the diameter and spacing must be small compared to a wavelength of, say, green light, which makes the manufacturer's task a difficult one.

It is amusing that one popular "explanation" of how polarizers work is exactly wrong. We have in mind the much-quoted picket-fence analogy, which runs as follows: "Consider a long taut rope that is passed between two pickets on a picket fence. Now shake one end of the rope. If you shake it up and down, the rope slides freely vertically between the pickets and the waves travel freely through the fence. If you shake the rope horizontally, the pickets restrain it and no vibration travels through." Students of electricity and magnetism will recall that the strongest eddy currents are produced in conductors having the *lowest* resistance; accordingly, the energy losses are *largest* for such conductors. It is equally true that electric virbations *parallel* to the wires of the wire-grid polarizer produce the largest currents and hence suffer the largest losses. The rope-and-picket-fence analogy may be helpful to beginners, but it is wrong by just 90°!

H-SHEET: WORLD'S MOST POPULAR POLARIZER

The commonest type of polarizer is H-sheet, invented by E. H. Land in 1938. It may be regarded as a chemical version of the wire grid. Instead of long, thin wires, it employs long, thin molecules—long-chain polymeric molecules that contain many iodine atoms. These long, straight molecules are aligned almost perfectly parallel to one another, and because of the conductivity provided by the iodine atoms, they strongly absorb the electric vibration component parallel to the molecules. The component perpendicular to the molecules passes on through, with very little absorption.

Ingenious schemes are used for aligning the "conducting" molecules. The manufacture starts with a large transparent sheet of easily stretched and chemically reactive plastic—usually polyvinyl alcohol. The sheet is warmed and then quickly stretched to many times its original length; as it becomes longer, it also becomes narrower and thinner, which accentuates the stretching.

During the stretching operation most of the long polymeric molecules, originally having random orientations, become turned so as to have nearly the same direction, namely, the direction of the stretching force. The simple stretching operation, performed by hand or preferably by machine, aligns the billions of molecules. It is hard to conceive of a simpler or faster means of alignment. The process can be demonstrated with the aid of a broad strip of rubber. One lays some matches on the strip at various orientations in the neighborhood of 45°, then stretches the strip. The matches are observed to turn so as to lie more nearly parallel to the strip. The method is not quite perfect; for example, matches that are initially exactly perpendicular to the stretch direction do not turn at all.

As soon as the soft, pliable sheet has been stretched, it is cemented to a rigid sheet, for example a sheet of cellulose acetate, so that no unstretching can occur.

The next step is to dip the sheet into a liquid solution that is rich in iodine. Within seconds much iodine has diffused into the polyvinyl-alcohol layer and becomes affixed to the thin, nearly parallel molecules. Thus the iodine atoms themselves, imitating their host, tend to form long, thin "chains" which are too small to be seen under a microscope but can be examined by x-ray diffraction techniques.

The sheet is then washed, dried, and cut into pieces of any desired shape and size. Cover plates of glass can be provided to protect the plastic layers from scratches and fingerprints.

H-sheet is made routinely on large machines that can turn out thousands of square feet per day.

The long, thin iodine chains act like wires. They absorb electric vibrations parallel to their alignment axis and transmit freely the vibrations perpendicular to that axis. Thus the *transmission axis* is perpendicular to the *stretch direction*. Ideally, 50 percent of an incident beam's power is transmitted, or about 45 percent if one recognizes that some of the light is lost by reflection from the surfaces. In practice about 38 percent is transmitted by one popular brand (Polaroid HN-38). By allowing a larger amount of iodine to enter the sheet, the manufacturer can reduce the leakage of light of unwanted vibration-direction to an extremely small value, such as 0.001 percent of the incident light. Unfortunately the transmission of the *wanted* component is thereby reduced slightly. Polarizers designated as HN-32 transmit about 32 percent of the incident, unpolarized light, whereas HN-22 transmits about 22 percent. The latter type is virtually free of any leakage of visible light.

Figure 3-3 illustrates the degree of success achieved by typical

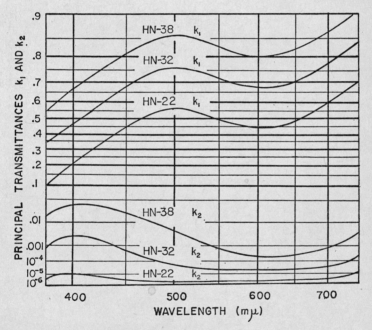

FIG. 3-3 Transmittance curves of various grades of H-sheet. The k_1 and k_2 curves represent the transmitted fractions of *wanted* and *unwanted* components respectively. (Ideally, $k_1 = 1.00$ and $k_2 = 0$.)

HN-38, HN-32, and HN-22 polarizers. The k_1 curves show the fraction of *wanted* vibration component that is transmitted, as a function of wavelength of the light; the k_2 curves show the fraction of *unwanted* component that is transmitted. The fractions differ by nearly six orders of magnitude, so a "logarithm of logarithm" scale is needed to show the values with adequate precision.

Small differences in the k_2 curves imply very noticeable differences in leakage of crossed polarizers. The difference between a good polarizer and a mediocre one depends on small differences of this type.

J-SHEET

So-called J-sheet, although superseded 25 years ago by H-sheet, is famous for having been the world's first sheet-type polarizer and the first polarizer that enjoyed widespread use by scientists and the public. When first invented (in 1928, by E. H. Land, then an undergraduate at Harvard College) it created a sensation: for better or worse, it made such a deep impression on optical men that many of them still regard *sheet polarizer* and *J-sheet* as synonymous terms. They fail to realize that further advances have been made and that J-sheet is no longer produced.

J-sheet employed actual crystals—millions of them. The individual crystal was dichroic, i.e., it absorbed light to different extents depending on the vibration direction. Land chose to use crystals of *herapathite,* a mineral exhibiting very large dichroism. The crystals have a very favorable shape—a needle shape—and thus can be aligned relatively easily by simple mechanical methods. When aligned, the tiny crystals "vote like one man": that is, they all agree upon absorbing one vibration component of the incident beam and transmitting the orthogonal component. Thus a large sheet of plastic containing millions of aligned herapathite crystals acts like one huge crystal. Indeed, it does even better than this: it acts like one crystal of great length and breadth but extremely small thickness. This is almost ideal since a small thickness, such as 0.005 inch, provides a good compromise between high absorption of the unwanted component and high transmission of the wanted component.

The main drawback to J-sheet was a slight haziness that it exhibited. Crystals of diameter larger than one wavelength of light tend to scatter the light somewhat. Yet to produce much smaller crystals is expensive. Persons who used J-sheet found the scattering property to be a handicap in certain applications, and they were delighted to shift to H-sheet when this became available in about 1938. H-sheet, being a molecular polarizer rather than a microcrystalline polarizer, has practically no haziness.

K-SHEET

K-sheet, invented by Land and H. G. Rogers the year after H-sheet was invented, is superior to H-sheet in one important respect: it can withstand high temperature without decomposing. A polarizer placed close to a powerful lamp becomes hot, and K-sheet is well suited to enduring the high temperature. Thus it is ideally suited for use in front of lamps used to project polarization-coded stereoscopic pictures. Also, it is ideal for use in automobile headlamps, when and if polarization-coded glare-suppression systems are adopted. (These applications of polarizers are discussed in Chapter 10.) K-sheet is more expensive than H-sheet and hence is not as widely used.

K-sheet, like H-sheet, is made from polyvinyl alcohol. However, instead of *adding* atoms to the sheet, the manufacturer takes atoms away. By employing a hydrogen chloride catalyst and a high-temperature oven, he removes $2N$ hydrogen atoms and N oxygen atoms, leaving a different type of polymeric molecule, called polyvinylene. To align the polyvinylene molecules, the sheet is stretched unidirectionally.

HR-SHEET

When placed in a beam of near-infrared radiation, H-sheet shows little absorption and hence produces little polarization. The same is true of K-sheet. But by combining the techniques used in making these two types, a polarizer (called HR-sheet) can be made that exhibits strong absorption and strong polarization in the near infrared. The useful range of this polarizer is from 0.6 to 2.7 microns.

OTHER DICHROIC POLARIZERS

Many other kinds of dichroic sheet polarizers can be made. Some employ dyes—dyes that consist of long, thin molecules.

Also, several kinds of plastic sheet can be used—even familiar Cellophane. Usually one dissolves the dye in a hot liquid and then dips the plastic sheet into the solution, allowing the dye molecules to diffuse into the sheet. Ordinarily the dye molecules are aligned by means of a stretching operation.

It is possible to make a polarizer merely by rubbing a dye into the surface of a plate of glass; one employs a unidirectional rubbing motion that causes the dye molecules that adhere to the glass to have a common direction.

One large crystal of appropriate type can serve as a polarizer. The classic example is the tourmaline crystal. In 1815 the French scientist Biot discovered that tourmaline absorbs different polarization forms to different extents; that is, he discovered the phenomenon of dichroism. For many decades such crystals were used routinely in laboratories, but by present-day standards they perform poorly: the k_1 curve is not as high as one would like and the k_2 curve is not as low as one would like. In other words, the tourmaline polarizer is both murky and leaky. Also, it is small and expensive.

A single crystal of herapathite can be used, but it is too small. Attempts to grow larger crystals were continued spasmodically for a century, but without great success.

Even a liquid can serve as a dichroic polarizer, if it contains long, thin dye molecules and is streaming or flowing in such a manner that there is a shearing action. The shearing causes the molecules to become partially aligned, so that they collectively exhibit asymmetry of absorption. This is scarcely a practical method of making a polarizer, but it has been found to be useful in studies of the dye molecules themselves.

Can one produce a *circular* polarizer by any of the methods discussed above? Is there any kind of dye that absorbs a right-circularly polarized component more than a left-circularly polarized component? The answer is yes, but such materials are rare. Few, if any, polarizers suitable for use in the visual range have been made by this approach. In contrast, radio engineers could point to important uses made of helical antennas. Right-helical antennas exist, and also left-helical antennas; they behave very differently with respect to right- and left-circularly polarized radio waves.

4 *Polarizers of the Reflection Type*

ADVANTAGES AND DISADVANTAGES

Polarizers of the reflection type have several attractive features. They can be designed for use in any portion of the spectrum, and a single polarizer may perform reasonably uniformly throughout many octaves of spectral range (many factors of 2 in wavelength). Also, they are easy to make. Indeed, the world around us is filled with objects that act as partial polarizers of reflection type; nearly any very smooth, nonmetallic object tends to polarize light that strikes it obliquely. Examples are glossy paper, polished table-tops, wet roadways, and bald heads.

However, many polarizers of the reflection type perform poorly: the k_1 values are not great enough, and the k_2 values are not small enough. Some carefully designed reflection-type polarizers have high performance, but tend to be bulky and expensive, and they perform well only if mounted in a well-collimated beam, and at just the correct obliquity.

Consequently, reflection-type polarizers are seldom used to polarize light in the visual range; they are far outclassed here by H-sheet, for example. Only in certain parts of the infrared range, and perhaps in certain parts of the ultraviolet range, are reflection-type polarizers preëminent. Hence the brevity of this chapter.

PRINCIPLE OF OPERATION

One of the simplest reflection-type polarizers is a piece of plate glass that is mounted obliquely in the given beam of light. When the plate is mounted perpendicular to the beam, no polarization results; all components of the light are transmitted with equal efficiency (about 92 percent, ordinarily) and the transmitted beam is found to be unpolarized. About 8 percent is reflected, and this light too is unpolarized.

But when the plate is tilted, so that the symmetry of the reflection process is destroyed, interesting things happen, as the French scientist Malus discovered in 1808. The transmitted beam is found to be partially polarized, and the reflected beam even more so. The polarization forms of the two beams are orthogonal.

Why does oblique reflection produce polarization? The reason follows directly from the electromagnetic theory. In fact, Fig. 4-1

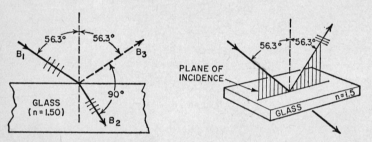

(Left) FIG. 4-1   Effect of a glass plate on a polarized beam B_1 that is incident at the polarizing angle and has a vibration direction parallel to the plane of the paper. The refracted beam B_2 is exactly perpendicular to the reflected beam B_3, and the *vibration direction* of B_2 is exactly *parallel* to beam B_3. Therefore, in fact, no energy is propagated in the B_3 direction.   (Right) FIG. 4-2   Effect of a glass plate on an *unpolarized* beam incident at the polarizing angle. The component having a vibration direction perpendicular to the plane of incidence is reflected moderately strongly ($\rho_{90} \cong 15$ percent) and the component vibrating parallel to that plane is reflected not at all ($\rho_0 \cong 0$ percent). Consequently the reflected beam is 100 percent polarized; it contains about 8 percent of the incident energy. The transmitted beam is only partially polarized and contains about 92 percent of the energy.

makes the reason clear without the need for any mathematics at all. It does so for one particular case—the case in which a beam B_1 is incident at 56.3° from the normal on a plate of glass of refractive index 1.50. We will suppose, first, that the beam is already 100 percent linearly polarized and that the electric vibration direction is parallel to the plane of the paper (and, of course, perpendicular to the axis of the beam). We now consider the refracted beam B_2 that enters the glass and beam B_3 that is reflected from the upper surface. The refracted beam has a somewhat steeper direction than B_1, as described by Snell's law; and the vibrations here are, of course, perpendicular to this steeper direction. The crucial fact is that, in the particular case at issue, *the vibrations in refracted beam B_2 are exactly parallel to the direction of the reflected beam B_3.* (That is why we picked the 56.3° angle of incidence.) And this in turn means that *the reflected beam B_3 cannot exist.* There is no such beam. No energy can flow in that direction. Why? Because, according to electromagnetic theory, light requires a transverse vibration; yet at the point where the light starts to enter the glass the vibrations happen to be exactly *parallel* to the direction B_3 and hence *have no component* that is transverse to B_3.

But suppose the incident beam has the orthogonal polarization form; that is, suppose the electric vibration is *perpendicular* to the plane of the paper. In this case the vibration direction of the reflected beam is perpendicular to the paper and perpendicular also to the direction B_3. Accordingly, transverse waves can and do travel in this direction (as well as along the refracted-beam direction).

Finally, suppose that the incident beam B_1 is unpolarized, as indicated in Fig. 4-2. How does the glass plate affect this beam? It "mentally" divides the beam into two components C_1 and C_2 vibrating, respectively, parallel to and perpendicular to the plane determined by the three rays, and (for reasons indicated above) it reflects the first component to "zero extent" and the second to a moderately large extent. The reflected beam thus consists solely of the second component, and hence exhibits 100 percent polarization. The transmitted beam also shows polarization, but to a much smaller extent and, of course, with the orthogonal vibration direction predominating.

Is the situation the same when a different angle of incidence is chosen? No. Neither component C_1 nor component C_2 produces a refracted beam having a vibration exactly parallel to the direction of reflected beam B_3. Consequently, each component delivers some power in the beam B_3 direction, and the degree of polarization in this beam is less than 100 percent.

The unique angle that yields a completely polarized reflected beam is known as Brewster's angle, or the *polarizing angle*. It is the only fact that one needs to know in order to produce a completely polarized reflected beam. Knowing the refractive index *n* of a given plate of glass, plastic, or the like, anyone familiar with Snell's law can compute the polarizing angle; it turns out to be *the angle that has n for its tangent*. Thus a plate of silver chloride ($n = 2.0$) has a polarizing angle of arc tan 2.0, or about 63°. Water has an index of 1.33; thus natural light that is reflected at arc tan 1.33, or about 53°, from the surface of a pond is completely polarized. These angles are measured from the normal to the surface.

With the aid of the wave equations of electromagnetic theory, one can make accurate predictions as to the reflectance (reflection factor) of a smooth dielectric plate when a linearly polarized beam of light strikes the plate obliquely. The wave equations involve boundary conditions; the surface of the plate is the boundary, and the refractive index is one of the conditions. By solving the pertinent equations one can calculate exactly how much light is reflected from a dielectric plate of any given index, for any given angle of incidence, and for any choice of vibration direction of the incident beam. When one examines the answers, he finds that they depend strongly on the vibration direction assumed.

Figure 4-3 summarizes the results for the two extreme choices of vibration direction in the incident beam, namely the cases in which the vibration direction is *parallel* to the plane of incidence and *perpendicular* to this plane. The symbols ρ_0 and ρ_{90} stand for the respective reflectances. Specifically, ρ_0 is the ratio of reflected energy to incident energy when the vibration direction of the incident beam is parallel to the plane of incidence. The quantity ρ_{90} is the corresponding ratio for the case in which the vibration direction is at 90° to this plane and thus is parallel to the surface of the plate. In the special case in which the angle of incidence is equal to the polarizing angle, the former reflectance is zero and the latter is fairly large, and hence the degree of polarization of the reflected beam is 100 percent. As one chooses angles of incidence that differ more and more from the polarizing angle, the ratio of the two reflectances approaches unity and the degree of polarization of the reflected beam approaches zero. Thus unpolarized light incident at grazing angle

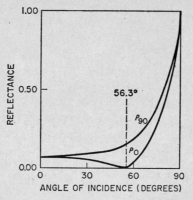

FIG. 4-3 Reflectance of a single surface of a dielectric plate in vacuum. The refractive index is assumed to be 1.50, and the incident beam is assumed to be linearly polarized with a vibration direction parallel to (0°) or perpendicular to (90°) the plane of incidence.

remains unpolarized; likewise unpolarized light incident along the normal remains unpolarized.

Thus far we have considered the reflection from just one surface of the plate. Plates possess two surfaces; consequently two reflection processes occur in series and the combined reflected beam is more intense. The polarizing angle remains unaffected, and when an unpolarized beam is incident at this angle the degree of polarization in the reflected beam remains 100 percent.

Although the light reflected from a glass plate may exhibit a high degree of polarization, the transmitted beam usually does not. When employing a single plate as a polarizer, one usually uses the reflected beam. The transmitted beam can be allowed to continue onward harmlessly or can be eliminated by applying black paint to the back surface of the glass (or by employing glass that is black throughout).

Since, typically, the refractive index changes only slightly with wavelength, the polarizing angle also changes only slightly with wavelength. Thus a reflection polarizer that is mounted at a suitable compromise angle (angle appropriate to some intermediate wavelength) may perform well throughout a broad spectral range, such as two or three octaves.

By the photon theory of light, the above discussion is wrong in a certain way. Each photon is either totally reflected or totally transmitted. For any one photon the reflectance is 0 or 1. No matter how a penny falls toward the floor, it comes to rest heads or tails. Similarly, no matter how the photon approaches the glass plate, and no matter what its original polarization, it must be entirely reflected or entirely transmitted. Only when millions of photons are involved and the average outcomes are measured do the figures agree closely with the predictions of the electromagnetic theory.

PILE-OF-PLATES POLARIZER

Usually one wants a polarizer that will deliver an energetic beam. When using a single-plate reflection polarizer, one finds that the reflected beam is weak; it contains only a small fraction of the incident beam's energy. Fortunately there is a simple way to increase the intensity: use many plates in series. For example, one may use a pile of 15 glass plates, e.g., 15 microscope slides. One arranges them all mutually parallel, with enough free space between neighboring plates so that no colored rings or other interference effects will arise. One then mounts the pile at the polarizing angle in the given beam. As more plates are added, the intensity of the reflected beam increases. However, diminishing returns soon set in because of absorption by the glass or scattering by dust on the surfaces.

The usual way of using a pile-of-plates polarizer is *in transmission*: the pile is placed in the optical apparatus in such a way that the transmitted beam is used and the reflected beam is

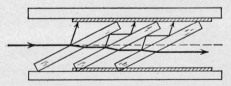

FIG. 4-4 Longitudinal section of a three-plate reflection-type polarizer used in transmission. Most of the (unwanted) reflected light eventually reaches the black-painted holder and is absorbed. The transmitted beam has the same direction as the incident beam, but the axis is displaced slightly.

discarded. Using the transmitted beam, the investigator does not have to rearrange the rest of his apparatus: the beam direction remains unchanged when the polarizer is inserted. However, the beam axis may be shifted laterally a small distance, as indicated in Fig. 4-4.

Usually 5 to 15 plates are used. The higher the index, the fewer plates are needed. Silver chloride and certain other materials well known for their transparency to infrared radiation have a high index (2.0 or greater); accordingly, six plates may suffice in an infrared polarizer. For best results the plates must be well polished, clean, dust-free, and free of strain-induced birefringence; birefringence leads to alterations in polarization form and hence to reduction in the degree of polarization, as explained in Chapter 6. Pile-of-plates polarizers consisting of only six plates of silver chloride can be designed to produce a transmitted beam of infrared light that is 99 percent polarized. Here too the equations of the electromagnetic theory permit one to predict the outcome accurately.

The unwanted component of the light is reflected back and

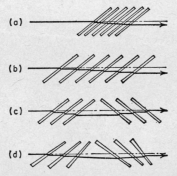

FIG. 4-5 Naive and sophisticated designs of a six-plate pile. (a) Naive design. (b) Design employing greater spacing of the plates, hence allowing better "ventilation" of the unwanted components. (c) Scheme employing two oppositely sloping sets of plates to improve the ventilation and bring the beam back to the original centerline. (d) Scheme employing groups of plates that are "wedged and fanned" so that those portions of the unwanted, multiply reflected components that leak through the system will follow a slightly different direction from the main transmitted beam and can be stopped by suitable diaphragms.

forth several times between the successive plates. Eventually most of it reaches the blackened walls of the holder and is absorbed, as indicated in Fig. 4-4. Some of it, however, may "leak through" and join the transmitted beam, thus reducing the degree of polarization. This difficulty becomes serious when the number of plates is large. Schemes shown in Fig. 4-5 may be used to give the unwanted light enough space to escape readily from the pile laterally, or to alter its direction slightly so that it follows a different path from the main transmitted beam.

5   *Polarizers of*
the Double-Refraction Type

In Chapters 3 and 4 we showed how to polarize light by employing asymmetry of absorption or asymmetry of reflection. Neither of these methods played any part in the original discovery of polarization.

Polarization was discovered with the aid of devices exhibiting asymmetry of refraction. It was discovered in 1690 by the Dutch scientist Huygens while he was studying some crystals of the doubly refracting mineral *calcite*. Accordingly, the calcite doubly refracting polarizer is the type mentioned first in most textbooks on light. But double refraction is a complicated subject; it requires some understanding of crystallography. Students (and textbook writers too) may become so immersed in the complexities of crystallography that they forget about light. Hence our postponement of the subject until this chapter. Even here we force ourselves to keep our attention on light, and we treat crystallography and crystal optics with a brevity that would dismay a crystallographer. (For a fuller account see Elizabeth A. Wood, *Crystals and Light: an Introduction to Optical Crystallography,* Momentum Book #5, D. Van Nostrand Company, Inc., Princeton, N. J., 1964.)

SIX BASIC FACTS OF CRYSTAL OPTICS

Six of the basic facts of crystal optics must be grasped if this chapter and the next are to be understood easily. These are as follows:

(1) When one aims a beam of light at a calcite crystal or other uniaxial doubly refracting crystal, one finds, in general, that inside the crystal there are *two beams* and that these are *surprisingly invariant in character.*

(2) Usually, one of these beams has a direction of energy flow that is oblique, rather than normal, to the wavefronts.

(3) Usually the two beams have different propagation speeds.

(4) Usually they have different propagation directions.

(5) Each beam is 100 percent polarized.

(6) The two polarization forms are orthogonal.

In Fig. 5-1 we attempt, not overly successfully, to indicate all these facts.

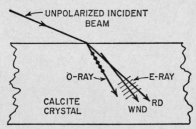

FIG. 5-1 Diagram designed to suggest, in exaggerated manner, six basic facts of crystal optics. The crystal is of calcite, and the optic axis is parallel to the plane of the paper. The diagram shows (1) the two refracted rays O and E, (2) two kinds of directions pertaining to the E-ray, namely, ray direction *RD* and wave-normal direction *WND*, (3) the difference in speed between the O-ray and the E-ray, (4) the difference in direction between the O-ray and the E-ray, (5) the polarization of the two rays, and (6) the opposite character of the two polarization forms, indicated by dots and hatch marks.

The facts are momentous. First, their importance to persons seeking to understand light is far-reaching; a full treatment of the effects of transparent crystals on light would fill an entire book. Second, their importance to persons trying to discover how crystals conduct sound, conduct heat, and diffract x-rays is perhaps equally important; many mysteries concerning crystals have been solved by seeing how the crystal affects a light beam. Double refraction was discovered in 1669 by the Danish scientist Bartholinus, and the repercussions are still echoing through the halls of science.

Let us now consider each of the six basic facts in more detail.

FIRST BASIC FACT: TWO INVARIANT BEAMS

It is indeed surprising that one incident beam can produce two refracted beams. It is even more surprising that these two are so "inflexible." No matter whether the incident beam is unpolarized or polarized, and no matter what variety of polarization forms is tried, the two refracted beams within a calcite crystal always have the same two propagation directions, the same two speeds, and the same two polarization forms. Using the language of photons, we say that the crystal compels each incident photon to choose between two "packages": (1) the fixed speed, direction, and polarization form of one refracted beam, or (2) the fixed speed, direction, and polarization form of the other refracted beam. No other speeds, propagation directions, or polarization forms are allowed as long as the propagation direction of the incident beam remains unchanged. The crystal's stubborn insistence on two and only two choices of polarization form is suggested by Fig. 5-2, which shows a great variety of polarization forms of incident beam, and shows no variety at all in the vibration directions of the two refracted beams (although in certain special cases one of the beams may be missing). The intensities and phases may change, but the vibration directions are invariant. The two beams reduce to one when the light travels within the crystal in the unique direction called the *optic axis*, discussed in a later paragraph.

SECOND BASIC FACT: OBLIQUE ENERGY FLOW

When light travels in vacuum, nothing very exciting happens: the energy is propagated in a direction that is exactly normal to the wavefronts. The same is true of light that travels in air, or

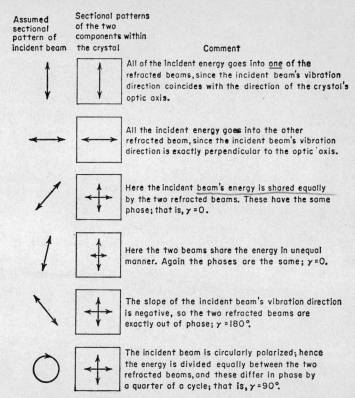

Assumed sectional pattern of incident beam	Sectional patterns of the two components within the crystal	Comment

All of the incident energy goes into one of the refracted beams, since the incident beam's vibration direction coincides with the direction of the crystal's optic axis.

All the incident energy goes into the other refracted beam, since the incident beam's vibration direction is exactly perpendicular to the optic axis.

Here the incident beam's energy is shared equally by the two refracted beams. These have the same phase; that is, $\gamma = 0$.

Here the two beams share the energy in unequal manner. Again the phases are the same; $\gamma = 0$.

The slope of the incident beam's vibration direction is negative, so the two refracted beams are exactly out of phase; $\gamma = 180°$.

The incident beam is circularly polarized; hence the energy is divided equally between the two refracted beams, and these differ in phase by a quarter of a cycle; that is, $\gamma = 90°$.

FIG. 5-2 Effect of aiming beams of various polarization forms at a calcite crystal. The incident beam is traveling horizontally straight toward the crystal and the observer. The crystal's optic axis is vertical. Note that the refracted beams have the *same* two polarization forms in every case. Nothing differs but their relative intensities and relative phases.

water, or glass, or any other isotropic material. (An isotropic material is one that produces identical effects on beams of light *regardless of the directions involved.*)

When light travels within a calcite crystal, however, the situation is very different. The direction of energy flow, called the ray direction (RD), often *differs* from the wavenormal direction (WND). It is the latter kind of direction (wavenormal direction) with which an optical designer deals. Many calculations are easier when he specifies this direction, rather than the ray direction. Likewise, he usually deals with a special kind of speed, namely, "speed measured along the normal." This is defined as $V_r \cos A$, where V_r is the true speed of energy flow, i.e., the speed measured along the ray direction, and A is the angle between wavenormal direction and ray direction. In the following pages we deal almost exclusively with wavenormal direction and "normal speed."

THIRD BASIC FACT:
DIFFERENT PROPAGATION SPEEDS

Here two comments are required. The first is that the *normal speed* of a refracted beam (either beam) depends only on that beam's vibration direction. It does not depend on the wavenormal direction, or on the ray direction either. This means that beams having the same vibration direction always have the same normal speed even when the wavenormal directions differ widely. And conversely, beams that have different vibration directions have different normal speeds (usually), even when the wavenormal directions are the same. To repeat: the crucial direction, as far as normal speed is concerned, is the vibration direction. (If light vibrations were longitudinal, there would be no parameter left to explain why two beams that follow the same path in calcite can differ in speed. The existence of double refraction is prime evidence that the vibrations are transverse.)

The second comment is that whenever there are two refracted beams within a uniaxial crystal, one of them always has the same invariant speed. In calcite, for example, the slower of the two beams always has the identical speed, namely, 1.808×10^8 m/sec, assuming that we are dealing with yellow light of wavelength 5893 angstroms. This beam is called the ordinary ray (O-ray). The other beam's normal speed varies: it depends on the particular direction of the vibration, and may be as small as 1.808×10^8 m/sec or as large as 2.017×10^8 m/sec, or anywhere in between.

Instead of dealing with speed, one may deal with refractive index. The index of the O-ray (the *ordinary* index, ω) is defined as the ratio of c (the speed of light in vacuum) to the normal speed of the O-ray. Its value is $(2.998 \times 10^8$ m/sec)/$(1.808 \times 10^8$ m/sec), or 1.658. It is sometimes called the major principal refractive index. The index of the E-ray (called the *extraordinary* index, ϵ') varies, depending on the vibration direction, and may be as large at 1.658 or as small as 1.486, or anywhere in between. The extreme value is called ϵ, the minor principal refractive index.

These two principal indices, called ω and ϵ (or n_O and n_E), are the key numbers that designers of calcite polarizers keep in mind. The virtue of calcite is that its two principal indices differ so widely. The difference is called the birefringence, and the symbol J or Δn is used. For calcite, $J \equiv |\epsilon - \omega| = |1.486 - 1.658| = 0.172$. Because $\epsilon < \omega$, calcite is said to be a negative crystal.

No other readily available, highly transparent crystal has such a large birefringence. Crystalline sodium nitrate ($NaNO_3$) has a birefringence $J = |1.336 - 1.587| = 0.251$, but large crystals are not found in nature and must be grown artificially. Certain special crystals used in infrared instrumentation have even greater birefringence.

FOURTH BASIC FACT: DIFFERENT PROPAGATION DIRECTIONS

In general, the two refracted beams have different wavenormal directions. But in certain cases these two become identical: This occurs when the incident beam happens to be perpendicular to the front face of the crystal. Both of the refracted beams' wavenormal directions are then perpendicular to this face, and hence are parallel to one another. But even in this special case, the two normal speeds differ and the two *ray* directions differ.

There is one very special situation, however, in which every possible simplification occurs: the wavenormal directions are alike, the normal speeds are alike, and the ray directions are alike. To be specific, there is one choice of crystal face such that a perpendicularly incident beam acts in the same simple way as it acts when entering glass: no double refraction occurs! This special direction of incident beam (and the exact opposite direction) is called the *optic axis* of the crystal. Finding this axis (i.e., *direction*) in a calcite crystal is difficult: a naturally

occurring calcite crystal has no face with the requisite orientation, so such a face must be prepared by grinding and polishing.
Calcite, having just one such axis, is called uniaxial. The same
applies to quartz, although here $\epsilon > \omega$ and the crystal is called
positive. Some crystals have two such axes, i.e., two such directions, and are called biaxial.

The reason the O-ray always has the same normal speed in a
given uniaxial crystal is that the vibration direction of this ray
is always perpendicular to the optic axis. For the E-ray the situation is very different: its vibration direction may be parallel to,
perpendicular to, or oblique to the optic axis, depending on the
crystal face used and the angle of incidence; when the vibration
direction is exactly *parallel* to the optic axis, the normal speed
has its extreme value, i.e., in calcite a maximum value and in
quartz a minimum value.

FIFTH BASIC FACT: 100 PERCENT POLARIZATION

The statement that each of the two refracted beams is 100
percent polarized is no rough approximation, no crude exaggera-
tion. In a good specimen of calcite, each refracted beam is linearly
polarized to the extent of at least 99.999 percent and perhaps
more. The remaining discrepancy, if any, is probably due to
imperfections in the measurement technique, not due to the
crystal. Where else can one find such a clean division into two
pure forms?

SIXTH BASIC FACT: ORTHOGONAL POLARIZATION
FORMS

This fact should evoke no surprise. Once we recognize that the
crystal divides the incident beam into two invariant polarization
forms, we should not be surprised that the forms are orthogonal.
Indeed, men with a flair for thermodynamics have proved that
the forms *must* be orthogonal: if they were not, the second law
of thermodynamics could be violated. A beam of light, like other
energy-containing systems, has an entropy value, and the rules
against entropy decrease would be violated if a crystal divided
an unpolarized beam (completely disorderly system) into two

similarly or nearly similarly polarized beams (more orderly system) traveling along the identical path.

When one thinks in terms of photons and quantization, one is quite prepared to take Facts 5 and 6 for granted. Quantization suggests clean categories and clean antitheses, so one is not surprised that the two refracted beams are ideally clean as regards degree of polarization and ideally antithetical (orthogonal) as regards form of polarization.

CONSTRUCTING A CALCITE POLARIZER: THE GLAN-FOUCAULT POLARIZER

It will now be clear that constructing a polarizer out of calcite is easy. Indeed, any smooth-faced calcite crystal *is* a polarizer. If one takes such a crystal and holds it at nearly any orientation in a very slender beam of light, the emerging light is found to be polarized. The crystal "can't help" dividing the incident beam into two polarized beams. When inside the crystal they travel in slightly different directions. They emerge at slightly different places and thus constitute two side-by-side beams of linearly polarized light with mutually perpendicular vibration directions.

But unless the incident beam is very slender or the crystal is very large, the two emerging beams may overlap a good deal. In the region of overlap there is no polarization, since the two orthogonal forms add together, and with no systematic phase relationship. Use of a big enough crystal avoids the problem, but large crystals are hard to find. Accordingly one seldom makes polarizers in this crude way.

An excellent design of calcite polarizer is that invented by Glan and Foucault, and shown in Fig. 5-3. Two pieces of calcite are used. Each is triangular in cross section, i.e., prism-shaped, with an apex angle θ of about 38.5°. Each has been cut (from a large, natural, oblique-faced crystal) so that the geometrical axis of the prism is parallel to the optic axis. The two pieces are mounted side by side, held so that there is a thin air gap between the respective hypotenuses, as shown in the figure. When a slender beam of unpolarized light strikes one prism along the normal to the face, it enters the prism and forms two beams there, namely, an O-ray and an E-ray. What happens when

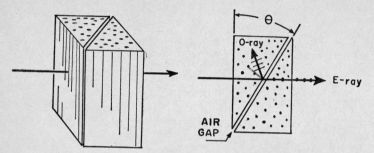

FIG. 5-3   The Glan-Foucault polarizer. The vertical hatch marks indicate the direction of the optic axis.

these strike the hypotenuse face? Are both transmitted? Or are both totally internally reflected? The inventors chose such an angle (for the hypotenuse) that one beam is totally internally reflected and the other is transmitted. The reflected beam strikes some black paint on the side of the first prism and is absorbed. The transmitted beam proceeds onward, into the air gap. As it does so it is bent sharply to one side, according to the usual rules of refraction; but it then encounters the second prism and is bent the other way by the same amount. Thus the final direction is parallel to the original direction. The degree of polarization is practically 100 percent.

The only interesting parameters of the crystal are the two principal indices ω and ϵ (1.658 and 1.486, respectively). Knowing these, and knowing Snell's law, anyone can compute in a few minutes a prism-apex angle that will cause one component of the light (the slow component—the O-ray) to be totally internally reflected while allowing the other (the fast one—the E-ray) to emerge. The size of the prisms is, of course, irrelevant to the polarizing capability; the degree of polarization is just as high whether the prism thickness is 1 mm or 100 mm.

Since calcite is transparent from about 2300 Å in the ultraviolet to about 5 microns in the infrared, and since the Glan-Foucault polarizer contains nothing but calcite, this polarizer can be used throughout a very wide spectral range. However, it performs properly only when the incident beam strikes the entrance face squarely and contains no rays making large angles with the normal—larger than about 7°. Dichroic polarizers avoid this limitation; a piece of HN-22 sheet, for example, produces a high degree of polarization even in rays incident at 30° from the normal.

If the two prisms of the Glan-Foucault polarizer are cemented together, thereby eliminating the air gap, the device is called a Glan-Thompson polarizer. This has an acceptance far exceeding 7°, but the cement that is usually employed is opaque to ultraviolet light.

NICOL POLARIZER

The most famous double-refraction-type polarizer is the one invented in 1828 by the Scottish physicist Nicol. For nearly a century the Nicol prism was the work-horse polarizer of optical laboratories the world over. Today it is outclassed by several other types and is mainly of historic interest. This fact is a boon to the present authors, since it justifies their omitting any detailed account of the design; the design is a complicated one, and a full description would fill several pages. Figure 5-4 suggests some key features and some of the complexities; for example, all of the faces of the polarizer are oblique, and the optic axis is oblique to every face.

The original crystal of calcite must be cut along a diagonal plane, then cemented together using a cement having a refractive index intermediate between calcite's ordinary and extraordinary indices. This assembly divides an incident beam into two refracted beams, totally internally reflects one (at the first interface between calcite and cement), and transmits the other. The time-honored choice of cement is Canada balsam, as its index is almost ideal, namely, 1.55.

The Nicol performs well, but has a limited spectral range because of absorption of ultraviolet light by the Canada balsam. There are several minor drawbacks also, including lateral displacement of the beam, and astigmatism due to the obliquity of the entrance and exit faces.

OTHER POLARIZERS OF DOUBLE-REFRACTION TYPE

Many other designs of calcite polarizer have been invented and put to use. Each can claim some advantage, such as higher major

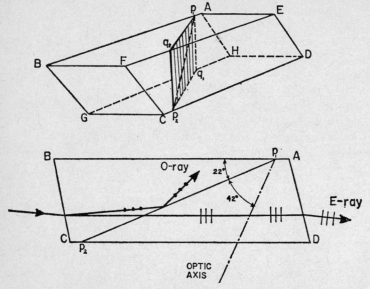

FIG. 5-4 The Nicol polarizer. The plane $p_1q_1p_2q_2$ is the plane where the calcite crystal has been cut in two and then cemented together again with Canada balsam. The four lateral faces are coated with black paint.

principal transmittance k_1, lower minor principal transmittance k_2, greater acceptance angle, smaller astigmatism, smaller over-all length, or greater spectral range.

The Wollaston polarizer, mentioned in Chapters 1 and 2 and illustrated in Figs. 2-4 and 5-5, is noteworthy in that it preserves

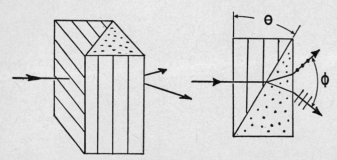

FIG. 5-5 The Wollaston polarizer. Both of the refracted beams emerge. Different apex angles θ may be used, each leading to a different total deviation angle ϕ.

both of the refracted beams but separates them widely enough so that each can be used separately. It is both a polarizer and a *polarizing beam-splitter*. It employs two calcite prisms cut so as to have mutually perpendicular directions of optic axis. The two prisms are cemented together.

The Rochon polarizer is generally similar to the Wollaston, except that the orientation is changed by 90° so that, in the first prism, the light travels parallel to the optic axis. One beam emerges from the polarizer in straight-ahead direction and the other in an oblique direction. Usually the latter beam is discarded.

SCATTERING-TYPE POLARIZERS

A polarizer designer may employ asymmetry of absorption, reflection, refraction, or scattering. Scattering-type polarizers usually perform poorly and are of little practical importance. But the principle of operation is worth considering, because in clear weather the earth's atmosphere acts as a vast scattering-type polarizer and causes the shorter wavelength (blue and ultraviolet) light from the sky to be polarized to some extent.

The asymmetry involved in polarization by the atmosphere is the asymmetry of the observer's location relative to the primary radiation. That is, the observer must gaze in a direction roughly perpendicular to the sun's rays. Let us consider a nearly horizontal sun's ray B, shown in Fig. 5-6. When this ray strikes a certain molecule M, it causes transverse vibrations in the electronic structure of the molecule, and these in turn produce secondary waves traveling in all directions. An observer off to one side, viewing the transverse, horizontal ray S_h, will receive only a vertical vibration. Any horizontal vibration that was perpendicular to his line-of-sight would be *parallel* to the primary ray B; but B contains no longitudinal vibration (according to electromagnetic theory) and hence cannot produce any such vibration. Thus ray S_h is polarized, linearly and vertically.

Likewise, an observer directly above or below molecule M can receive light of only a single vibration direction, namely, the direction that is perpendicular both to the primary ray and to his line-of-sight.

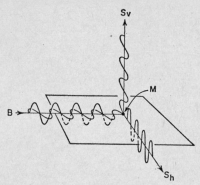

FIG. 5-6 Snapshot pattern of rays S_h and S_v scattered at 90° by air molecule *M* struck by a horizontal unpolarized beam *B*. These rays are highly polarized.

What happens when the observer looks straight toward the sun? He receives light of all vibration directions, or rather, of all that are perpendicular to his line of sight. None of these is forbidden, since none is parallel to the primary ray. Thus the light he sees is unpolarized, even though it reaches him from the same scattering molecule *M*. The degree of polarization of skylight is discussed in Chapter 10.

Scattering-type polarizers can be made, also, of turbid solutions or suspensions, provided that the scattering entities, i.e., the dissolved molecules or suspended microparticles, are small compared to the wavelength of light used. When these entities are too large, they may reflect more than they scatter, and the outcome is very different.

6   *Retarders*

RETARDERS AS POLARIZATION-FORM CONVERTERS

Retarders, or *wave plates,* are converters of polarization form. By employing the appropriate retarder, one can convert any given form into any other form. And the conversion process is almost 100 percent efficient.

A main use of retarders is in the production of out-of-the-ordinary polarization forms. An experimenter first produces a linearly polarized beam (by using H-sheet, for example), then interposes the appropriate retarder in the beam to achieve the desired polarization form. Circular polarization and elliptical polarization are usually produced in this way, i.e., in two steps.

The converse use is important too. Given some unusual polarization form that one wishes to analyze and identify, one tries to find what retarder, introduced in the beam at what orientation, will convert the beam to familiar linear polarization form.

The typical retarder suffers from chromatism: it produces changes of different magnitude in beams made up of different wavelengths. When placed in a beam of white light, it produces a grand mixture of magnitudes-of-change. To simplify the exposition presented in the next few pages, we make the assumption that the light beams are monochromatic.

PRINCIPLE OF OPERATION: THE 180° RETARDER

The principle of operation of a retarder is easily stated: it divides an incident, polarized beam into two components, changes the phase of one relative to the other, then combines them again. Because of the phase shift, the combined beam has, as we shall see, a different polarization form. Since the key step is the introduction of a relative phase shift, some people call the device a *phase shifter*.

A typical retarder consists of a thin crystalline plate with its optic axis parallel to the plane of the plate, as indicated in Fig. 6-1. Let us assume that the plate is of calcite and rests on one edge, with the optic axis horizontal. Let us assume, next, that a linearly polarized beam is incident on the plate horizontally from the left, as indicated in the figure, and let us consider the two refracted beams that exist within the plate. Both are linearly polarized, as we saw in Chapter 5. One has a vertical vibration direction. (Why? Because one of the beams must be the O-ray, and an O-ray's vibrations are always perpendicular to the optic

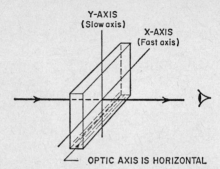

FIG. 6-1 Calcite 180° retarder, with optic axis horizontal, mounted squarely in a beam that is incident horizontally from the left.

axis.) The other has a horizontal vibration direction, because it must be orthogonal to the first.

The crucial fact is that the two beams travel at different speeds, so that when they emerge, one has been "retarded," or slowed, compared to the other. Of course, both rays are considerably slowed when they travel in the calcite. Both are retarded. But we are interested only in *relative* retardation. We ask only: How much more was the O-ray slowed down than the E-ray? We use the word *retardance* to mean this difference.

In calcite it is the O-ray that travels more slowly. So in the present case it is the beam with the *vertical* vibration direction that is slowed, and this vertical direction is called the slow axis *S*. The direction of vibration of the faster-traveling beam is called the fast axis *F*.

If the slow beam were retarded by just one wavelength, i.e., just one full cycle, no one could know that the retarder had done anything: One wave is like the next. So if the two beams, on emerging, join together exactly one cycle, or two cycles, or three, etc., out of step, no significant change has been made. No conversion of polarization form has been accomplished.

But suppose the slow ray emerges just one-half wavelength behind the fast ray. That is, suppose it has been retarded by just 180° in phase. Is the polarization form of the emerging beam different from that of the incident beam? The answer is yes. Fig. 6-2 shows why, for the case in which the incident beam is linearly polarized at 45°. This incident beam is equivalent to two components (two simple harmonic motions) the vibrations of which are parallel to the X- and Y-axes, respectively, and are in the same phase, as indicated in Fig. 6-2a. But if the vertical com-

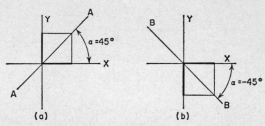

(a)   (b)

FIG. 6-2 Conversion of +45° polarization form to the −45° form by means of a 180° retarder oriented with the fast axis horizontal. The X- and Y-components of the incident beam AA, shown in (a), have the same phase, but the components of the emerging beam BB, shown in (b), are 180° out of phase, and accordingly the resultant is at −45°.

ponent is retarded by 180°, it must be indicated by a line having the opposite direction, i.e., downward, rather than upward, as suggested by Fig. 6-2b; and the resultant then lies on a line at *minus* 45°. The 180° retarder has converted +45° linear polarization to −45° polarization.

The 180° retarder, or *half-wave plate,* is famous for its ability to alter polarization form symmetrically with respect to the retarder's axes, here assumed to be parallel to the X- and Y-axes. An incident beam linearly polarized at +78°, say, is converted to one at −78°; and similarly for any other choice of angle α of vibration direction of the incident beam. It is always as if the 180° retarder had reflected the incident beam's vibration direction in a mirror, the mirror always being parallel to the retarder's fast axis (or, equally well, to its slow axis).

The 180° retarder can be used to convert right-circularly polarized light into left-circularly polarized light, or vice versa. Also, it can be used to convert right-elliptically polarized light into left-elliptically polarized light and at the same time "reflect" the major axis of the ellipse about the fast (or slow) axis.

In practice, 180° retarders are not made of calcite. Calcite's birefringence is large, and accordingly the plate would have to be extremely thin, and would be fragile. Usually quartz or mica is used. Their birefringence is much smaller, so the thickness that gives 180° of retardance is much larger and more convenient than with calcite. Grinding and polishing quartz pieces to proper thickness requires specialized equipment, but cleaving mica is simple; usually one cleaves dozens of thin sheets and picks out those that happen to have the right retardance. Any sheet that succeeds in converting +45° linear polarization to the −45° form fills the bill: its retardance is 180°—unless perchance the thickness is much too great and the retardance is (180° + 360°), or (180° + 720°), etc. It is also possible to make retarders from large sheets of organic plastic, such as polyvinyl alcohol, that consist mainly of long, thin polymeric molecules; when the sheets are warmed and then stretched unidirectionally they become permanently birefringent.

THE 90° RETARDER

The 90° retarder, or *quarter-wave plate,* is another useful device. It can convert linearly polarized light into circularly polarized light, and vice versa. Also it can convert linear forms into elliptical forms, and vice versa. Here too one could use a plate of calcite, but quartz, mica, or birefringent sheets of organic polymeric plastic are usually used.

No diagram is needed to explain the action of a 90° retarder on a beam of linearly, horizontally polarized light. What the retarder does is divide the incident beam into two linearly polarized components and retard one of them by a *quarter* of a cycle. Thus the emerging beam's sectional pattern represents the combination of two simple harmonic motions (along mutually perpendicular axes) that are 90° out of phase with one another. The pattern of the combined beam is not a straight line, but an ellipse. The ellipse becomes a circle if the retarder is turned so that its axes are at +45° and −45° to the vibration direction of the incident beam, i.e., so that the two components have equal amplitude. The handedness of the circle is left or right depending on whether the fast axis of the retarder is at −45° or +45°, respectively.

When a 90° retarder is inserted in a circularly polarized beam, it divides this into two components that are *already* out-of-phase by 90° and then adds 90° to this phase difference (or subtracts 90° from it), and in either case the emerging beam is linearly polarized. Knowing the orientation of the retarder and having determined the vibration direction of the emerging beam, one can deduce whether the incident beam was right- or left-circularly polarized. (The deduction can be made almost instantly, at a glance, when one employs the short-cut "Poincaré sphere" method discussed in Chapter 8. Also, many people can see *with the naked eye* whether the sense of circular polarization is right or left; the method is indicated in Chapter 10.)

Retarders need not be made of doubly refracting materials; the retardance can be provided by asymmetry of reflection. A prime example is the Fresnel rhomb illustrated in Fig. 6-3a. It is made

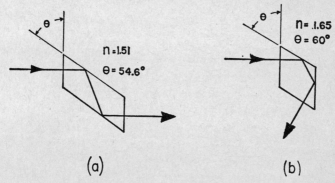

FIG. 6-3 Designs of (a) the Fresnel rhomb and (b) the Mooney rhomb.

of glass, and is designed to produce total internal reflection of an incident beam *twice,* at the oblique faces. Each reflection retards one component by 45° relative to the other; thus the cumulative retardance is 90°. An attractive feature is that the retardance is almost constant throughout a wide range of wavelengths. (As indicated in a later paragraph, this is not true of typical retarders made of birefringent material.)

The Fresnel rhomb has one big limitation: the need for the incident beam to be well collimated and almost exactly perpendicular to the entrance face. The Mooney rhomb shown in Fig. 6-3b permits much greater latitude, but produces an emergent beam having a "diagonally backward" direction.

Warning: When using a linear retarder, one is dealing with two kinds of angles. One angle, sometimes called ρ, specifies the orientation of the fast axis of the retarder; it is a real, "spatial" angle. The other angle, called δ, specifies the retardance; it has no relation to real space, but specifies, in a "mathematical" space, the phase lag of the slower beam. Consider, for example, a quarter-wave plate whose fast axis is at 17°. Here $\rho = 17°$ and $\delta = 90°$.

The CP-HN polarizer: Instead of using a linear polarizer and an entirely separate 90° retarder to produce circularly polarized light, one may cement the two devices together—in correct orientation. The CP-HN circular polarizer produced commercially is made in this way. A large sheet of 90° retarder consisting of stretched polyvinyl alcohol is cemented, at proper 45° orientation, to a large sheet of H-type polarizer. The combined sheet may then be cut into large numbers of pieces of suitable size for laboratory use. In using this laminated, two-layer type of circular polarizer, one must place the device "correct side front" in the given beam. If it is wrong side around, the last layer the light traverses is the H-sheet layer and accordingly the emerging light is *linearly* polarized.

CIRCULAR RETARDERS

Certain crystals, when suitably oriented in a beam, resolve it into two components that are polarized circularly, not linearly. One component is right circular and the other left circular. One component travels more slowly than the other. Here, then, one has the opportunity of making a circular retarder. (The word *circular* here has nothing to do with the actual shape of the retarder, but merely specifies the class of polarization forms that occur within the crystal.) Elliptical retarders exist also, i.e., retarders that produce refracted beams whose polarization form is elliptical. As indicated in Chapter 10, some liquids, too, act as circular retarders. Under some circumstances, gases may also.

Materials that act as circular retarders are called *optically active,* or *rotators,* and are said to exhibit circular birefringence. Circular birefringence is always a consequence of a handedness, or helicity, in the molecular or crystalline structure of the material in question. Crystalline quartz has such a handedness, which, in any individual crystal, is either of right or left type. Such a crystal exhibits circular, elliptical, or linear birefringence with respect to beams traveling parallel to, oblique to, or perpendicular to the optic axis, respectively.

RETARDANCE vs BIREFRINGENCE

The retardance of a plate depends, obviously, on the birefringence. But the two quantities should not be confused. Birefringence is an *intensive* quantity; it is the difference between the two principal refractive indices, as explained in Chapter 5; it remains constant no matter what the size or orientation of the plate. It is, in short, a constant of the material.

Retardance is an *extensive* quantity. Double the thickness of a plate, and the retardance is doubled. The retardance, expressed in cycles, of a calcite plate cut parallel to the optic axis is given by Jt/λ, where J is the birefringence of the material, t is the thickness of the plate, and λ is the wavelength of light used. If the plate is cut perpendicular to the optic axis, J must be replaced by zero: there is no retardance. If the plate is cut obliquely, J must be replaced by some intermediate number, sometimes called the plano-birefringence.

EFFECT OF A RETARDER ON AN
UNPOLARIZED BEAM

The over-all effect of a retarder on an unpolarized beam can be stated in one word: none. This is true even though the retarder does indeed resolve the beam into two polarized components. It is true because each component has no definable average phase; hence there is no definable average phase relationship between them, and the very concept of retardance is inapplicable. When the two components join together (when they emerge from the retarder) the combination is just as chaotic as the incident beam was. No one sectional pattern is predominant: the degree of polarization is zero.

If the incident light is partially polarized, the retarder does

have an effect. To compute this, one mentally divides the incident beam into a polarized part and an unpolarized part and
then treats each of these separately by the methods described
previously in this chapter.

EFFECT OF A RETARDER ON
POLARIZED WHITE LIGHT

As stated at the outset of this chapter, a retarder of any common birefringent material has a retardance that varies considerably with wavelength; that is, the retarder is chromatic. The
retardance, expressed in degrees or cycles, varies approximately
as the reciprocal of wavelength and thus is 30 to 50 percent
greater for blue light than for red.

This circumstance opens the door to a flood of beautiful and
sometimes useful color effects. To produce such effects, one usually employs one retarder and two linear polarizers, placed in a
beam of white light. The light strikes first a polarizer, then the
retarder, and then the second polarizer. The first polarizer
polarizes all components of the beam (all wavelengths, all colors)
identically. But the retarder does *not* treat all wavelengths identically: having different retardance for different wavelengths, it
produces a variety of different polarization forms. If it is oriented
so as to transform green light, say, to the polarization form that
the second polarizer absorbs completely, no green light will
emerge from that polarizer. Other portions of the spectrum are
transformed differently and are *not* entirely absorbed. The upshot
is that the emerging light is rich in blue and red, but lacking in
green. Thus it appears purple. When the retarder is turned to
other angles, other colors are absorbed entirely, and some green
light, for example, passes through. A great variety of vivid colors
can be produced.

To produce a nonchromatic, or *achromatic,* retarder is very
difficult, if one insists on using a doubly-refracting material. One
must find, or produce, a material whose birefringence J is directly proportional to wavelength, so that the retardance itself,
Jt/λ, will be independent of wavelength. No fully satisfactory
solution to this problem has been found.

EIGENVECTOR OF A RETARDER

For a given retarder oriented in a given way, one can find two kinds of polarized beams (two polarization forms) that will pass through it without change, i.e., so that the emerging beam will have the same polarization form as the incident beam. Let us consider, for example, the calcite retarder shown in Fig. 6-1. Obviously, if the incident beam is polarized exactly vertically, the emergent beam is polarized vertically also, for reasons made clear by Fig. 5-2. Likewise a horizontally polarized incident beam emerges horizontally polarized.

The two polarization forms that pass through a retarder unchanged are called *eigenvectors* of the retarder. They constitute a kind of shorthand description of the retarder in terms of *what it cannot do,* i.e., the two polarization forms that it cannot alter. The one associated with faster travel within the retarder, i.e., the fast eigenvector, is used routinely in the Poincaré sphere method of specifying a retarder and predicting its effect on a given beam of polarized light, as indicated in Chapter 8. The eigenvectors of a retarder are *constants,* not of the retarder in general, but of the *retarder in the given orientation.* If the retarder is turned to a new angle, two new eigenvectors apply.

The eigenvectors of a circularly birefringent retarder are simply right- and left-circular polarization: either right- or left-circularly polarized light passes through the retarder without change in polarization form. Elliptical retarders, too, have eigenvectors. These are elliptical polarization forms having opposite handedness and mutually perpendicular directions of major axis and the *same* ellipticity. Turning a circular retarder to a new angle does not change the eigenvectors, since a circle looks the same regardless of orientation.

SEVERAL RETARDERS IN SERIES

What happens when several retarders are placed in series? Is the over-all effect of the "train" equivalent to that of one single retarder? The answer must be yes if the light beam is monochromatic. We have already seen that any one polarization form can be converted to any other specified form by means of a single retarder. In other words, there is no transformation that cannot be accomplished by *one* retarder. Trains of retarders can produce a wide variety of transformations of polarization form, but an experimenter can always find a *single* retarder that can accomplish the same thing. The Poincaré sphere method of Chapter 8 enables one to find this equivalent single retarder immediately.

PSEUDO-DEPOLARIZERS

When one places several birefringent retarders in series at various angles to one another, the effect on an incident beam of linearly polarized *white* light is very complicated, in that components having different wavelengths emerge with different polarization forms. The polarization status of the emerging beam is so complicated that the beam might easily be mistaken for an unpolarized beam. Accordingly, trains of retarders are often used to "depolarize" light. One must place "depolarize" in quotation marks, since, in fact, light of any one wavelength remains polarized. It is merely the combined effect of all wavelengths that, in various practical circumstances, roughly simulates depolarization. Accordingly, such a train is a pseudo-depolarizer, not a real depolarizer. It is perhaps worth emphasizing that retarders can accomplish any desired transformation of polarization form, but cannot abolish it. They can neither create polarization nor destroy it. To produce a real depolarizer would be just as difficult as to produce a machine that would shuffle cards "perfectly."

7   *Conventional Algebra of Polarizers*

The authors have had a difficult time writing this chapter. Their hearts were not in it. The subject is not exciting, and the

PLATE I   H-type polarizers almost big enough to hide behind. They were cut from a sheet several hundred feet long. When crossed at 90° they produce almost perfect extinction. (Photograph courtesy of the Polaroid Corporation)

PLATE II   Black-and-white photograph of the photoelastic pattern produced when a plastic disk, compressed between two metal bars, is
placed between crossed polarizers and is illuminated from behind. The
actual pattern is highly colored. (Photograph courtesy of the Polaroid
Corporation)

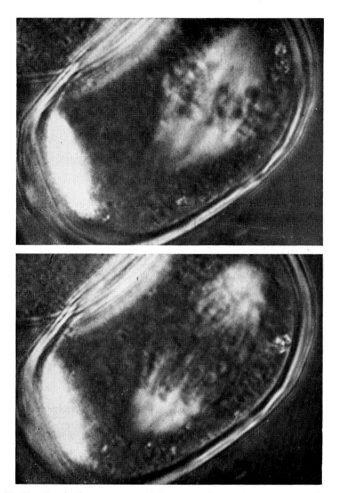

PLATE III   Single frames printed from a 16-mm time-lapse motion picture taken with the aid of a high-power polarizing microscope designed
to reveal very small amounts of retardance in objects only a few microns in diameter. In the upper frame the bundle of chromosomal fibers
is in process of division; in the lower frame the division is complete.
Such fibers are invisible when examined with an ordinary microscope,
but because they are birefringent they can be made to show up clearly
in a microscope employing crossed polarizers. (Photograph courtesy of
Dr. S. Inoué of the Dartmouth Medical School; reproduced with permission of Academic Press, Inc.)

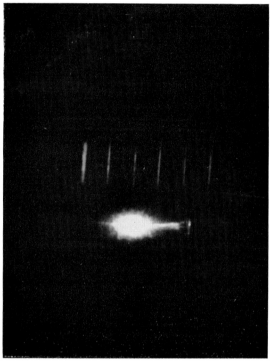

PLATE IV Head-on view of blindingly bright polarized light: synchrotron radiation, emitted by orbiting electrons in the Cambridge Electron
Accelerator. With the aid of an expendable mirror situated only 2 inches
from the orbit and a 100-power telescope situated 50 feet away behind
a thick shielding wall of concrete, observers obtain such a view of the
self-luminous cross section of the beam of 6-billion-electron-volt electrons. The horizontal line to the right of the elliptical cross section is
produced at the end of the acceleration interval when the electrons
begin to lose energy and follow a path of gradually decreasing radius.
The light is about 85 percent linearly polarized, with electric vibration
direction horizontal. The vertical lines of the reference scale are 1
centimeter apart. (Photograph by W. A. Shurcliff)

methods have no glamorous future, since newer methods have been invented. When complications arise, the newer methods should be used. They are faster and, as explained in the following two chapters, provide more insight into the processes concerned. In the present chapter, therefore, the conventional methods are treated as briefly as possible.

PURPOSE OF THE ALGEBRA

An investigator who inserts one or more polarizers and/or retarders in a beam of light usually wants to know two things about the light that emerges: the polarization form, and the power relative to that of the incident beam. The conventional algebra of polarized light provides answers easily enough in simple situations; for each type of situation, a well-established formula is available. The formulas are based on the key parameters of the individual polarizers and retarders and on the orientations (azimuths, or angles) at which they are mounted in the beam.

EFFECT OF ONE POLARIZER

The key parameters of an ordinary linear polarizer are the principal transmittances k_1 and k_2, defined in Chapter 3. Briefly, k_1 is the transmittance which the polarizer exhibits when inserted in a linearly polarized beam and turned to the azimuth that maximizes the intensity of the emerging beam, and k_2 is the transmittance found when the polarizer is turned so as to minimize the intensity. Ideally, $k_1 = 1$, meaning that at one extreme orientation the polarizer transmits all the power in the incident polarized beam, and ideally $k_2 = 0$, meaning that, when turned to the other extreme orientation, the polarizer blocks the beam entirely.

The transmittances mentioned above are intensity (or power) transmittances: they are ratios of emergent intensity to incident intensity. Sometimes it is more convenient to deal with amplitude transmittances, which, in accordance with the basic laws of electromagnetic theory indicated in Chapter 2, are merely the square roots of the respective intensity transmittances. Ordinarily, the unadorned term "transmittance" means intensity transmittance.

For a typical polarizer, and indeed even for a so-called neutral polarizer, the principal transmittances vary slightly with wavelength. In the following paragraphs this complication is ignored; the simplifying assumption is made that the light is monochromatic and hence only two parameters (k_1 and k_2, or alternatively $\sqrt{k_1}$ and $\sqrt{k_2}$) are required to specify a given polarizer fully.

The orientation of a linear polarizer is usually specified in terms of the azimuth, or angle θ, of the transmission axis. This angle is usually measured from the horizontal; but if a linearly polarized beam with fixed direction of electric vibration is incident on the polarizer, θ may be measured from this vibration direction. A counterclockwise rotation of the polarizer is called positive.

A basic question is: When a linear polarizer is inserted in a 100 percent horizontally polarized beam and is turned to some azimuth θ, what is the actual transmittance T? That is, what is the actual ratio of emergent intensity to incident intensity? If θ happens to be $0°$, T has the value k_1, by definition. If θ is $90°$, $T = k_2$. When θ has some intermediate value, the answer is a little less simple. In the special case in which the polarizer is ideal, so that $k_1 = 1$ and $k_2 = 0$, the answer is: $T = (\cos \theta)^2$, as the French scientist Malus found in 1808. A present-day physicist would derive the formula by invoking the electromagnetic theory, undreamed of in 1808; he would reason that the amplitude A of the incident beam must be projected onto the transmission axis, to yield the transmitted amplitude, and this must be squared to give the transmitted intensity. Thus he would arrive at the expression $(A \cos \theta)^2$. The ratio of this to the incident intensity A^2 is $\cos^2 \theta$.

In dealing with the general case in which k_2 as well as k_1 is significant, a physicist would project the amplitude A of the polarized incident beam onto the transmission axis and also the orthogonal axis (absorption axis). The two projections have magnitudes $A \cos \theta$ and $A \sin \theta$. He would multiply these by $\sqrt{k_1}$ and $\sqrt{k_2}$ respectively, to take into account the amplitude transmittances associated with the two axes. To find the resultant of the two transmitted (coherent) amplitudes, he would take the square root of the sum of the squares; and to find the intensity he would square this, obtaining simply:

$$A^2 k_1 \cos^2 \theta + A^2 k_2 \sin^2 \theta.$$

Dividing by A^2, he would obtain the actual transmittance:

$$T = k_1 \cos^2 \theta + k_2 \sin^2 \theta.$$

The emerging beam is linearly polarized with the vibration direction parallel to the resultant of the two transmitted amplitudes. When k_1 and k_2 have the ideal values 1 and 0, the formula reduces to $T = \cos^2 \theta$ and the vibration direction of the emerging beam is identical to that of the transmission axis of the polarizer.

If the incident beam is circularly or elliptically polarized, the procedure is similar, except that before the two emergent components are combined, account must be taken of the phase relationship between them, in the manner indicated in Chapter 2.

When the incident beam is unpolarized, a different formula applies. Its derivation is almost obvious. The incident beam is regarded as being the sum of two *incoherent* components, one vibrating parallel to the transmission axis and the other parallel to the absorption axis. Each has an intensity equal to half that of the incident beam as a whole, and, by virtue of the definitions of k_1 and k_2, the transmitted intensities of the components are $\frac{1}{2}k_1$ and $\frac{1}{2}k_2$. Thus the total transmittance, called k_t, is simply: $k_t = \frac{1}{2}(k_1 + k_2)$. The transmitted beam is partially linearly polarized, and the dominant vibration direction is parallel to the transmission axis. The procedure does *not* involve amplitudes; this is true whenever the beams to be combined are incoherent.

When the incident light is partially polarized, the polarized and unpolarized fractions must be considered separately, and the outcomes combined. The unpolarized fraction has no systematic phase, so the combining process is accomplished merely by adding intensities.

TWO POLARIZERS IN SERIES

The formulas that apply to a pair of identical linear polarizers are nearly as simple. When an incident, unpolarized light beam of unit intensity strikes the first polarizer, the beam may be regarded as divided into two mutually incoherent compo-

nents, each of intensity $\frac{1}{2}$, that have vibration directions parallel
to the transmission axis and absorption axis of this polarizer.
The transmissions of the components are governed by the transmittances k_1 and k_2 of the first polarizer, and the resulting intensities are $\frac{1}{2}k_1$ and $\frac{1}{2}k_2$. The second polarizer may be regarded
as acting on each of these two components separately. If per-
chance it has the same orientation as the first polarizer, the
same respective transmittances k_1 and k_2 apply, and the total
intensity emerging from the second polarizer is $\frac{1}{2}k_1^2 + \frac{1}{2}k_2^2$. If
the second polarizer has a crossed, or $90°$, orientation relative to
the first, the transmittances k_2 and k_1 apply, in that (reversed)
order, and the total intensity emerging is $\frac{1}{2}k_1 k_2 + \frac{1}{2}k_2 k_1$, which
reduces to $k_1 k_2$.

The quantity $\frac{1}{2}(k_1^2 + k_2^2)$ is known as H_0, the transmittance
of the parallel pair, and the quantity $k_1 k_2$ is called H_{90}, the
transmittance of the crossed pair. It is easily shown that for any
intermediate angle-of-crossing θ, the intensity emerging has an
intermediate value that varies between the largest (H_0) and the
smallest (H_{90}) in simple, cosine-squared manner:

$$H_\theta = H_{90} + (H_0 - H_{90}) \cos^2 \theta.$$

This is an outstandingly useful formula. If $H_{90} \ll H_0$, the H_{90}
term may be omitted and the formula reduces to $H_0 \cos^2 \theta$.

TRAIN OF POLARIZERS AND RETARDERS

The effect of an individual retarder on a polarized beam has
been discussed at length in Chapter 6. A linear retarder, for example, divides a linearly polarized incident beam into two coherent components vibrating in mutually perpendicular directions that correspond to the fast and slow axes of the retarder,
and retards the phase of the latter relative to the former. The
polarization form of the emerging beam can be determined by
the graphical or mathematical means indicated in Chapter 2.
The intensity is identical to that of the incident beam, except
for the small reflection losses. An elliptical retarder acts similarly,
except that the polarization forms of the two components are
elliptical, rather than linear; the elliptical forms are simply the
eigenvectors of the retarder, as explained in Chapter 6.

When a beam goes through a series of polarizers and retarders, the outcome can be predicted with the aid of conventional algebra. The procedure, however, is long and dull. The intensity and polarization form of the light emerging from each element (each polarizer or retarder) must be determined explicitly, the output of one element constituting the input to the next. At all times close attention must be given to the directions of the axes of the elements, the vibration directions and amplitudes of the two component beams, and the phase relationships. If the given problem is altered by a change in polarization form of the initial beam or by a change in orientation of the first element, the entire computation must be performed afresh. All such worries and repetitions are avoided when the newer methods, described in Chapters 8 and 9, are used.

OTHER PARAMETERS OF POLARIZERS

Sometimes authors describe polarizers in terms other than those used in previous paragraphs. Instead of specifying the principal transmittances k_1 and k_2, they may specify the *principal optical densities* d_1 and d_2, defined as $\log_{10}(1/k_1)$ and $\log_{10}(1/k_2)$ respectively. When dealing with a dichroic polarizer, they may state the *dichroic ratio* R_d, defined as d_2/d_1, instead of the transmittance ratio R_t, defined as k_1/k_2. Dichroic ratio is a particularly valuable parameter of a dichroic polarizer because, to a first approximation, it is independent of the thickness. That is, polarizers made in different thickness, or with different concentrations of the given dichroic absorber, tend to have about the same dichroic ratio. This ratio is an intensive parameter and thus serves as a factor of merit of families of dichroic polarizers made from different materials (different *dichromophores*). A value of 30 or more is considered good; a value below 5 or 10 is usually considered unsatisfactory.

Just as a retarder has a retardance δ, so a polarizer has a *polarizance P*. This is defined as the degree of polarization produced when the polarizer is inserted in an unpolarized beam. Thus $P = (k_1 - k_2)/(k_1 + k_2)$. Since it is a constant of the polarizer, P remains unchanged when the polarization form or degree of polarization of the incident beam is changed.

Polarizers, like retarders, have eigenvectors. The eigenvectors of a polarizer are defined as the polarization forms (of incident beam) that remain unchanged when the polarizer is inserted at some given orientation in the beam. For any polarizer, two such polarization forms can be found. For example, the eigenvectors of a linear polarizer oriented with its transmission axis horizontal and its absorption axis vertical consist of (1) horizontally linearly polarized light and (2) vertically linearly polarized light. When the polarizer is turned through a given angle, it has two new eigenvectors; the eigenvectors always pertain to a *given polarizer in a given orientation*. The eigenvector associated with high transmittance is called the major eigenvector, and the one corresponding to low transmittance is the minor eigenvector. An ideal polarizer has a k_2 value of 0; therefore such a polarizer transmits no light specified by the minor eigenvector.

When a polarizer is to be used in a beam of unpolarized white light, and the transmitted intensity is to be judged by the human eye, the polarizer's weighted-average total transmittance throughout the visual range (about 400 to 700 millimicrons or 4000 Å to 7000 Å in wavelength) is of interest. Weighting is necessary in order to take account of the eye's reduced response near the ends of the visual range and to take account of the spectral energy distribution of the light source itself. The resulting quantity is called k_v. Another interesting quantity, called k_{vx}, is the weighted transmittance of a *crossed pair* of identical polarizers. The k_v values of common varieties of H-sheet are "built into the name"; thus HN-38, HN-32, and HN-22 have k_v values of about 0.38, 0.32, and 0.22 respectively. The k_{vx} values are about 0.000 5, 0.000 05, and 0.000 005 respectively.

8 *Modern Descriptions of Polarized Light*

POINCARÉ SPHERE

The Poincaré sphere is a kind of map. More specifically, it is a unit-radius spherical surface each point of which signifies a different polarization form. Any problem involving the effect of a retarder on a monochromatic, polarized beam is solved by "navigating" on the sphere. The process is simple, since the navigation is always along the arc of a circle. Some authors describe the navigation process as *rotating the sphere;* but this is a misnomer, since the actual procedure is to draw the arc and leave the sphere fixed.

In some optics laboratories, actual spheres made of wood or plastic are used. Usually the sphere is painted white, and lines of latitude and longitude are indicated in black. The arc that applies in a given problem can then be drawn with a colored crayon. Sometimes it is sufficient to draw a perspective sketch of the sphere and indicate the arc, also in perspective. A person familiar with the use of the sphere can solve simple problems in rough, or qualitative, manner in his head merely by shutting his eyes and visualizing the sphere. When greater accuracy is required, spherical trigonometry may be used.

Figure 8-1 shows how to map the various polarization forms onto the sphere. The north and south poles stand for left and right circular polarization, respectively. Every point on the equator represents a linear polarization form; each point implies a different vibration direction. Points between the equator and the south pole represent right elliptical polarization. Perhaps the most important point is the point *H* on the equator: it represents light that is linearly polarized with the electric vibration direction horizontal. Latitude and longitude are reckoned from

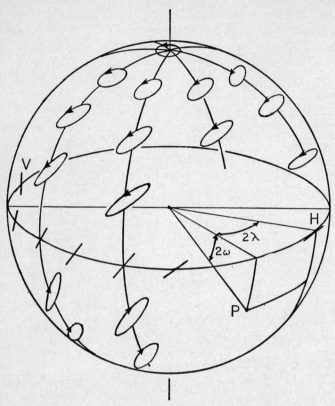

FIG. 8-1 Perspective view of the Poincaré sphere, showing the general significance of the different areas and the specification of the general point *P* in terms of the angle 2λ measured clockwise from *H* and the angle 2ω measured downward from the equator.

here. The point *V*, diametrically opposite *H*, represents *vertically* linearly polarized light. Any two diametrically opposite points represent an orthogonal pair of polarization forms. Figure 8-1 is not drawn perfectly; to show ellipses having many ellipticities and many azimuths was difficult enough for the draftsman, and to show them in perspective on the curving side of a sphere was even more difficult. The drawing succeeds, at least, in showing the general correlation between polarization form and position on the sphere.

Applications to Retarders: The Poincaré sphere is a "natural" for retarders. It provides a quick method of finding the effect of *any* retarder on *any* monochromatic beam of completely polarized light. The effect is found by marking the point P that designates the polarization form of the incident beam, marking the point R that designates the fast eigenvector of the retarder, and then drawing the appropriate arc. The axis of the arc is the radius vector from the center of the sphere to point R, and the starting point of the arc is point P. The length of the arc, in degrees, is simply the retardance δ of the retarder. The arc is always to be drawn in clockwise manner as judged by an observer situated outside the sphere, on an extension of the radius vector through R. The final point of the arc is the answer: it designates the polarization form of the light emerging from the retarder. In a typical problem, the quantities R, P, and δ are given. So the problem practically solves itself.

The method is universally applicable. The retarder may be of linear, circular, or elliptical type, and the polarization of the incident beam may have any form whatsoever. In every case the problem is solved in identical manner. The Poincaré sphere, being spherically symmetrical, plays no favorites. Circular or elliptical forms can be handled just as easily as linear forms.

Let us consider a simple example. Suppose that a beam linearly polarized at 45°, as indicated by point P in Fig. 8-2, encounters a 90° linear retarder the fast axis (fast eigenvector) of which is at 22.5° as indicated by point R. What is the polarization form of the emerging beam? The answer is found by drawing an arc the axis of which is the radius vector through R; the arc must start at point P; its length must equal the retardance, namely, 90°; accordingly the final point is P'. This is the answer. Since it lies in the "northern" hemisphere, it represents left elliptical polarization. Since it lies on the meridian through R and the north pole, it implies that the major axis of the ellipse has the same slope as that implied by R, namely, 22.5°.

The simplicity of the method is striking. No attention has to be paid to amplitudes, phases, or intensities, since all the wisdom needed is incorporated in the sphere. In fact, the sphere may be thought of as an analog computer, so ingeniously contrived by

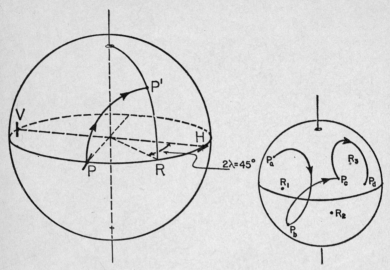

(Left) FIG. 8-2 Use of the Poincaré sphere in predicting the effect of a 90° linear retarder R (with fast axis at 22.5°) on a linearly polarized beam P (with vibration direction at 45°). The polarization form of the emerging beam is indicated by point P'. **(Right) FIG. 8-3** Use of the Poincaré sphere in finding the result of interposing a train of three elliptical retarders in a beam of elliptically polarized light. Points R_1, R_2, R_3 indicate the fast eigenvectors of the three retarders. Point P_a describes the polarization of the incident beam, and P_d describes the polarization of the finally emerging beam. The lengths of the three arcs correspond to the three retardances.

Poincaré that the entire process reduces to marking two points and drawing one arc.

Often the final point of the arc constitutes an adequate answer *as is:* it does not need to be interpreted in terms of ellipticity, handedness, and slope of the major axis. This is especially true when there are several retarders in series; in such a problem the end point of one arc serves as the starting point for the next. Figure 8-3 illustrates the method for a case that would be very hard to handle by conventional methods. Using the Poincaré sphere to deal with several retarders in series is like using a slide rule to carry out several multiplications in series: the intermediate values may be forgotten; only the final answer has to be read off and recorded.

When it is necessary to state explicitly the ellipticity, slope, and handedness represented by the end point of the arc, the following formulas are used. They refer to λ, which is half the longitude, and to ω, which is half the latitude; these angles are measured from the point H. The angle λ is positive if it is a *clockwise* angle as in Fig. 8-1; the angle ω is positive if it is measured *downward* as in that same figure. The formulas, which follow directly from Poincaré's definitions of the significance of a general point on the sphere, are as follows:

The ellipticity $b/a = \tan |\omega|$

The angle α of major axis of ellipse $= \lambda$

The handedness is left or right for 2ω negative or positive, respectively.

Obviously, λ and ω have quite different meanings here from those specified in Chapter 2.

The Poincaré sphere provides immediate answers to many basic questions concerning retarders. For example: Why does a circular retarder always convert one linear polarization form into another linear form? Because the fast eigenvector of the retarder is represented by one of the poles of the sphere, and any arc that is concentric with the polar axis and starts on the equator must continue on the equator. Why does a 90° linear retarder convert circularly polarized light into linearly polarized light? Because any 90° arc that starts at a pole and is drawn about a horizontal axis must end on the equator. Why does a 180° linear retarder produce the same effect whether its *fast* axis has a given azimuth or its *slow* axis has that same azimuth? Because the same end point applies whether the 180° arc is drawn about a given radius vector or about the diametrically opposed radius vector. Why is any combination of retarders equivalent to a single retarder? Because any combination merely converts the initial point, via several steps, to the final point, and it is always possible to find a *single* arc that does the same thing.

Applications to Polarizers: The Poincaré sphere is also useful in connection with problems involving polarizers: it can be used in predicting the effects of ideal polarizers on incident beams of polarized light. A single example will suffice. Let P_1 represent the polarization form of the incident, unit-intensity beam, and let P_2 represent the major eigenvector of the polarizer. What is the intensity of the emerging beam? The answer is found by following one brief instruction: Draw a great-circle arc between P_1 and P_2, find the length of the arc in degrees, take half this angle, look up the cosine of this half-angle, and square it. Thus if the arc has a length of 60°, the answer is $(\cos 30°)^2$, or 0.25.

Again the universality of the method is striking. The same procedure is used whether the incident beam is polarized linearly or elliptically and whether the polarizer is of linear or elliptical type. The sphere puts all forms of polarization on a par with one another.

To anyone familiar with the Poincaré sphere, the impossibility of describing unpolarized light by means of a plane diagram is particularly obvious: any truly "fair" description must give equal weight to every location on the sphere.

Some physicists have tried to extend the usefulness of the Poincaré sphere to problems involving partially polarized light. They propose that the sphere be regarded as a solid, or true, sphere, rather than a spherical surface. By introducing a number of additional definitions, they have succeeded in applying the sphere to a somewhat wider range of problems. Their methods will not be described here, since a better method of dealing with partially polarized and unpolarized beams is available; it employs the Stokes vector, considered in the next section.

STOKES VECTOR

The Stokes vector consists of a set of numbers—and no diagram. It applies equally well to polarized light, partially polarized light, and unpolarized light. Invented in 1852 by the British physicist G. G. Stokes, it provides the simplest possible method of predicting the result of adding two incoherent beams. (As explained in Chapter 2, two beams are called incoherent if there is no systematic phase relationship between them.) The method is purely numerical; it consists merely of adding the corresponding numbers from each set.

More important, the Stokes vector provides a simple numerical method of predicting how a beam is affected by the insertion of a polarizer or a retarder. The specification of the emerging beam is found by multiplying the set (Stokes vector) representing the incident beam by the set (matrix) representing the polarizer or retarder.

The set comprising the Stokes vector contains four numbers. For example, a unit-intensity unpolarized beam is specified by the set (1, 0, 0, 0); a horizontally linearly polarized beam is specified by (1, 1, 0, 0); a beam polarized linearly at 45° is specified by (1, 0, 1, 0); a right-circularly polarized beam is described by (1, 0, 0, 1). The orthogonal polarization forms are represented by vectors in which the crucial "+1" has been replaced by "−1". Thus the *vertical* linear form is (1, −1, 0, 0), the −45° linear form is (1, 0, −1, 0), and the *left* circular form is (1, 0, 0, −1). As should now be apparent, the second, third, and fourth numbers of a set indicate the "preferences" for horizontal linear form as compared to vertical form, for +45° as compared to −45°, and for right circular form as compared to left circular form. The range extends from +1 to −1, if the beam is of unit intensity, i.e., if it is *normalized*.

Actually, the Stokes vector is a column vector and hence should be written vertically, thus:

$$\begin{bmatrix} 1 \\ 0 \\ 0 \\ 0 \end{bmatrix}.$$

But often, to save space, it is written horizontally. When actually put to use in conjunction with matrices, in the manner indicated in the following chapter, it is written vertically.

The four numbers, or *Stokes parameters,* are called *I, M, C,* and *S,* respectively. The first number of the set denotes the intensity of the beam. Usually the incident beam is assumed to have unit intensity, so the first number is 1. Ordinarily the intensity decreases when the beam passes through a polarizer, so the first number pertaining to the emerging beam is less than 1. Its range is from 1 to 0.

When the given beam is completely polarized, the square of the intensity number equals the sum of the squares of the other numbers. That is, $I^2 = M^2 + C^2 + S^2$. When the beam is only partially polarized, the following inequality holds: $I^2 > M^2 + C^2 + S^2$, and the degree of polarization is given by

$$\frac{(M^2 + C^2 + S^2)^{1/2}}{(I)}.$$

How are the four numbers I, M, C, S found, for a given beam? If the point P_1 that represents the beam on the Poincaré sphere is known, the procedure is simple. First, a set of right-handed Cartesian coordinates is assigned to the sphere, in the manner indicated in Fig. 8-4. The origin is at the center of the sphere,

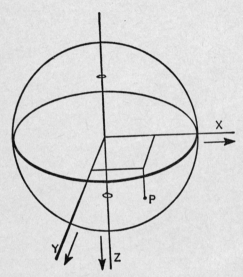

FIG. 8-4 Assignment of a right-handed Cartesian coordinate system to the Poincaré sphere, to permit finding the second, third, and fourth Stokes parameters of a given polarized beam represented by the point P.

the X-axis passes through the point H, the Z-axis extends straight downward through the south pole, and the Y-axis extends "toward the observer," through the point corresponding to 45° polarization. Then the X, Y, and Z coordinates of P_1 are the values of M, C, and S. The intensity, I, is specified separately, and is often assumed to be unity. If the latitude 2ω and the longitude 2λ of P_1 are known, X, Y, and Z may be found from the trigonometric relations:

$$M \equiv X = \cos 2\omega \cos 2\lambda$$
$$C \equiv Y = \cos 2\omega \sin 2\lambda$$
$$S \equiv Z = \sin 2\omega.$$

It is also possible to express I, M, C, and S in terms of the horizontal and vertical components a_x and a_y of the instantaneous electric displacement and in terms of the phase angle γ between these. The four pertinent expressions are as follows (the angular brackets indicate time averages):

$$I = \langle a_x{}^2 + a_y{}^2 \rangle$$
$$M = \langle a_x{}^2 - a_y{}^2 \rangle$$
$$C = \langle 2a_x a_y \cos \gamma \rangle$$
$$S = \langle 2a_x a_y \sin \gamma \rangle.$$

The validity of the first expression is obvious: the sum $(a_x{}^2 + a_y{}^2)$ is simply A^2, the square of the amplitude, and thus corresponds to intensity. The second expression reduces to $+1$ when the electric vibration is horizontal and the beam is of unit intensity, and reduces to -1 when the vibration direction is vertical; it becomes 0 when the beam is circularly polarized, elliptically polarized with major axis at $\pm 45°$, or unpolarized. Thus it does indeed indicate the preference for horizontal vibration as compared to vertical vibration. Similar reasoning shows that, as required, the third expression represents preference for $+45°$ vibration and the fourth expression represents preference for right circular polarization.

Happily, most of the Stokes vectors that are commonly required have been tabulated and published.*

Application: The result of adding any two incoherent beams is found by adding the two pertinent Stokes vectors. Consider, for example, the addition of a horizontally polarized beam of intensity *unity* and a right circularly polarized beam of intensity *three*. The pertinent vectors are $(1, 1, 0, 0)$ and $(3, 0, 0, 3)$. The sum, obtained by adding the correspondingly situated num-

---

* See, for example, W. A. Shurcliff, *Polarized Light: Production and Use*, Harvard University Press, 1962.

bers, is (4, 1, 0, 3). This resulting vector signifies that the intensity of the resulting beam is 4, since the first number is 4. The elliptical sectional pattern bears much resemblance to a circle, because the magnitude of the last number is large, and the handedness of the ellipse is right, because the last number is positive. The ellipse is more nearly horizontal than vertical, because the second number is positive. The degree of polarization is

$$\frac{(1^2 + 0^2 + 3^2)^{1/2}}{4},$$

or 79 percent.

Another example is the addition of two beams that are polarized horizontally and vertically, respectively, are incoherent, and are of unit intensity. Their vectors are (1, 1, 0, 0) and (1, −1, 0, 0), and the sum is (2, 0, 0, 0). This represents an unpolarized beam, since the last three numbers are zero. Would this same result apply if the two beams were coherent? Definitely not. If they were coherent and had the same phase, they would combine to form a beam linearly polarized at 45°. If they were coherent and differed in phase by 90°, they would combine to form a circularly polarized beam.

It is important to remember that Stokes vectors may be added only when the beams concerned are incoherent. In interference experiments, several beams are derived from the same part of a single source; such beams are usually coherent, and accordingly the addition of the Stokes vectors can lead to an utterly false result.

The main use of the Stokes vector is in predicting the result of inserting polarizers and retarders in a given beam. This is an important subject, and it is dealt with in the next chapter. Using vectors to represent light beams and matrices to represent polarizers and retarders, an investigator can solve very complicated problems quickly, accurately, and without need for deep thought.

JONES VECTOR

The Jones vector, invented in 1941 by a 25-year-old American physicist, R. Clark Jones, is superior to the Stokes vector in some ways and inferior in others. It is superior in that it is applicable to the addition of coherent beams; also, it is more compact. It is inferior in that it cannot cope with unpolarized or partially polarized light, and in that it employs complex numbers.

The Jones vector is a two-element column vector. Each element describes one component of the electric vibration (at the given location). The first (upper) element indicates the amplitude and phase of the X-component, and the second (lower) element does the same for the Y-component.

What is the Jones vector of a horizontally linearly polarized beam? The X-component may be written $A_x e^{i\phi_x}$, as indicated in Chapter 2. The Y-component is zero. Thus the complete vector is $\begin{bmatrix} A_x e^{i\phi_x} \\ 0 \end{bmatrix}$. Vertically polarized light has a vertical component but no horizontal component; consequently its vector is $\begin{bmatrix} 0 \\ A_y e^{i\phi_y} \end{bmatrix}$.

A simplification, called *normalization*, is usually made. Both elements are divided by whatever number (real or complex) produces the greatest simplification and also makes the sum of the squares of the elements 1.0. For example, the first Jones vector presented above is divided by $A_x e^{i\phi_x}$, and becomes $\begin{bmatrix} 1 \\ 0 \end{bmatrix}$; this is the normalized Jones vector of horizontally linearly polarized light.

A note of warning is needed here. In the process of normalizing certain vectors, some of the information as to phase is eliminated, and accordingly the relationship between various normalized vectors may be incorrect as far as phase is concerned. In most problems, only one beam is involved; hence absolute phase is not at issue, and the normalized vector can be used. But if several beams are involved, it is safest to go back and use the original, full vectors.

A beam linearly polarized at $+45°$ has X and Y components that are equal; the normalized vector is $\begin{bmatrix} 1/\sqrt{2} \\ 1/\sqrt{2} \end{bmatrix}$, usually written as $\frac{1}{\sqrt{2}} \begin{bmatrix} 1 \\ 1 \end{bmatrix}$. Right and left circular forms have components that are equal in magnitude, but are out of phase by $90°$, that is, out of phase in the same way that i and 1 are out of phase. The normalized vectors are

$$\frac{1}{\sqrt{2}}\begin{bmatrix} -i \\ 1 \end{bmatrix} \quad \text{and} \quad \frac{1}{\sqrt{2}}\begin{bmatrix} i \\ 1 \end{bmatrix}.$$

The intensity of the beam is proportional to the sum of the squares of the magnitudes of the elements. In other words, the magnitudes of the elements correspond to amplitudes, not intensities. If each element of a vector is multiplied by 4, the intensity is increased by a factor of 16.

Applications: One simple use of the Jones vector is in predicting the result of adding two coherent beams. Consider a unit-intensity, horizontally polarized beam and a 16-times-as-intense vertically polarized beam. The beams are assumed to be coherent and to have the same phase. The individual vectors are $\begin{bmatrix} 1 \\ 0 \end{bmatrix}$ and $\begin{bmatrix} 0 \\ 4 \end{bmatrix}$. The result of adding the two given beams is found by adding the vectors. The sum is $\begin{bmatrix} 1 \\ 4 \end{bmatrix}$. This implies a beam linearly polarized at an angle given by arc tan $(4/1)$; that is, 76°. The intensity is given by $(1)^2 + (4)^2$, or 17.

Or consider the right- and left-circularly polarized beams the full Jones vectors of which are $\frac{1}{\sqrt{2}}\begin{bmatrix} -i \\ 1 \end{bmatrix}$ and $\frac{1}{\sqrt{2}}\begin{bmatrix} i \\ 1 \end{bmatrix}$. Again the result of combining the two (assuming they are coherent) is found by adding the vectors. The sum is $\frac{1}{\sqrt{2}}\begin{bmatrix} 0 \\ 2 \end{bmatrix}$ or, more simply, $\begin{bmatrix} 0 \\ \sqrt{2} \end{bmatrix}$. Clearly this represents a beam that is vertically polarized and has an intensity of $(\sqrt{2})^2$, or 2.

The main use of the Jones vector is in computing the effect of inserting polarizers and retarders in a given polarized beam. As indicated in the following chapter, the effect is computed by multiplying the Jones vector of the incident beam by the matrices that represent the polarizers and retarders. The result of the multiplications is the Jones vector of the emerging beam.

CURRENT ATTEMPTS TO FATHOM
POLARIZED LIGHT

It is becoming increasingly clear that light and matter belong, ultimately, to the same family. The family has no name, but is sometimes referred to as the set of fundamental particles. It includes more than 30 members—listed, for example, by Frisch and Thorndike in Momentum Book #1, *Elementary Particles*. It includes the well-known particles *electrons, protons,* and *neutrons,* and also various muons, pions, kaons, and so-ons. Every one of these particles has energy, momentum, spin, upper limit c on speed, the capability of exhibiting wave-like properties, the capability of interacting with other members of the family, and the capability of being suddenly created or suddenly destroyed. The photon has every one of these properties and must therefore be recognized as a member in good standing of the same family.

The photon's claim to membership is clinched by the observation that its spin (or angular momentum) is simply related to that of the electron. The electron has angular momentum of magnitude $h/4\pi$, and a circularly polarized photon has *exactly twice* this angular momentum—an arrangement that is highly appropriate in that when an electron reverses its spin direction while emitting a photon, the photon can carry away exactly the necessary amount of angular momentum so that this quantity is conserved. "Spin-wise," the photon is as closely related to the electron as a key to a lock. Actually, the exact 2-to-1 match of photon spin applies not only to the electron, but also to most other matter-particles. It is noteworthy, and indeed almost incredible, that the match is valid whether the energy of the photon is great or small. Photon energies vary over a majestic range, but the spin of the photon is as unchanging as the Rock of Gibraltar.

Further evidence of kinship of photons and matter-particles is the recently discovered *similarity of function* of the photon and the pion. Each is now believed to be a "force-carrier." The electric force between two charged bodies may be said to stem from an exchange of photons between them; likewise the nuclear force between two neutrons, say, may be attributed to an exchange of pions. In each case the exchange is fleeting and not directly observable; consequently the photons or pions involved are called virtual. It is presumed that gravitational forces likewise can be explained by the exchange of a particle (called the *graviton,* although never observed as yet). Perhaps there are other kinds of exchange-type particles. The photon, then, is just one of an elite group of three or more force-carriers, collectively called the *elementary bosons*.

Because of the kinship between the photon and matter-particles, physicists are now convinced that an excellent way to learn about light is to study its interactions with other particles. In short, stop thinking about *what light is* and concentrate on *what it does*. The way to learn the essence of a five-dollar bill is not to look at it, but to take it to a supermarket and spend it. Today's physicists take the logical-positivist view that questions as to *what a thing is* have little or no meaning; what counts is *what it does*.

Light interacts with other particles in an especially varied and revealing manner if the energy of the photon is of the order of millions or billions of electron volts. Such photons can create nearly all known types of particles, and the interactions can be observed *individually* by means of bubble chambers and certain other huge detectors that clutter the halls of high-energy laboratories. After analyzing many such interactions, theoretical physicists try to discover rules by which they can predict what will happen when a given type of photon strikes a given kind of matter. Unfortunately, it appears that predictions of micro-behavior are applicable only statistically. The most that physicists can hope to discover are the probabilities that apply, i.e., the cross sections, distribution curves, occupancy numbers, or expectation values—quantities foreign to most optical men of a generation ago. The fruits of the physicists' labor will presumably be a set of mathematical expressions that are verifiable statistically rather than in individual events, and will have more appeal to a young mathematician than to an elderly physicist. The days of the comfortable physical model (the vibrating rope, the elastic solid) are drawing to a close.

Tomorrow's description of polarized light will be couched in the same language used for particles in general and bosons in particular, i.e., the language called the linear field theory of quantum electrodynamics. The present authors will say nothing about these subjects for the best of all reasons: ignorance. But the following summary may be suggestive: The photon is a boson that can have any energy from zero to infinity and any linear momentum from zero to infinity. When it is traveling in empty space, the upper limit on speed is c and the lower limit also is c. The photon exhibits wave-like properties when passing through a diffraction grating and under various other circumstances. It has infinite lifetime when traveling in empty space, but can interact with any electrically charged matter-particle that it encounters. It enjoys participating in force-carrying exchanges. Photons are created when matter-particles undergo annihilation, and under appropriate circumstances a single high-energy photon can disappear, leaving two oppositely charged electrons. The spin of a circularly polarized photon is one unit, the unit being $h/2\pi$. Right- and left-circularly polarized photons have spins of $+1$ unit and -1 unit along the direction of motion; linearly polarized light may be regarded as a coherent combination of right and left circular forms, and unpolarized light as an *incoherent* combination of such forms. A compact way of specifying the polarization in terms useful to high-energy physicists is by means of vectors such as the Stokes vector.

The link between light and matter is, of course, a two-way street. To learn about light, a physicist must analyze matter; but it is equally true that to learn about matter he must study up on light. Both fields are growing in importance; each enriches the other.

9   *Short-Cut Calculus*

GOAL OF THE NEWER METHODS

Here we describe two short-cut methods of solving problems involving polarizers and retarders. The methods are "preprogrammed" in the sense that all the necessary logic has been built into them by the inventors. The user has little to do other than to carry out the arithmetic, following a fixed procedure.

The goal of the new methods is to reduce the task of solving a given problem to a *routine multiplication*. An analogy will make this clear. Consider four ordinary (nonpolarizing) glass filters A, B, C, and D, each of which has a known transmittance for sodium light; let the four transmittances be called F_a, F_b, F_c, and F_d. Now suppose that all four filters are placed in series in a beam of intensity I_0. The intensity I_e of the emerging beam is computed by multiplying I_0 by the product of the four transmittances, that is, by obtaining the product $F_a F_b F_c F_d I_0$. If the transmittances are 0.1, 0.2, 0.3, and 0.4, their product is 0.0024, and so $I_e = (0.0024)I_0$.

Can this same procedure—simple multiplication by *numbers* —be used when polarizers and retarders are involved? Unfortunately not. A polarizer has two transmittances, not just one; also, it may belong to any of several classes (linear, circular, or elliptical), may have either handedness, and may have any orientation. A retarder, too, is specified by several constants, not just one. Thus the simple multiplication of numbers is entirely inadequate here.

However, two nearly-as-simple procedures have been discovered. They entail multiplication of matrices, rather than of numbers. A matrix, being a set of several numbers, has "room" for all the necessary constants for describing a given polarizer or retarder in a given orientation.

The procedure is as follows: Each polarizer and each retarder is specified by a matrix; then, in place of the simple transmittances F_a, F_b, etc., the matrices M_a, M_b, etc., are used; finally, in place of arithmetic multiplication *matrix multiplication* is used. The method is described in detail on a following page.

These more versatile methods were discovered in the late 1930's and early 1940's by Professor Hans Mueller of Massachusetts Institute of Technology and R. Clark Jones of the Polaroid Corporation. It seems strange that the discoveries did not occur generations earlier, and it is sad to contemplate that, for so many years, optical men solved their problems involving polarizers and retarders by the old, cumbersome methods.

The two new methods, known as the Mueller calculus and the Jones calculus, complement one another. Each works well in some situations and not in others. In general, where one fails, the other succeeds. The Mueller calculus, being more widely applicable, is discussed first.

MUELLER CALCULUS

The Mueller calculus employs matrices and vectors. The matrices, called Mueller matrices, are symbolized by M_a, M_b, etc.; each specifies a given polarizer or retarder *in a given orientation*. (Change the orientation, and you must use a different matrix!) The vectors are the Stokes vectors of Chapter 8. The vectors that represent the incident and emerging beams may be called V_i and V_e respectively.

Let us see how the M's and V's are used in a problem involving just one polarizer. One simple rule suffices: To find how the incident beam V_i is affected by the given polarizer M, merely multiply V_i by M. The product is the answer: it is a vector V_e that gives the full story on the emerging beam: the intensity, degree of polarization, and form of polarization. In matrix notation the rule is expressed thus:

$$[V_e] = [M][V_i].$$

The square brackets serve as reminders that matrices and vectors —not simple numbers (scalars)—are involved.

An example will show how the multiplication is carried out. Only one bit of matrix algebra is required. In many problems, most of the numbers are zero, and the multiplication "telescopes" to almost nothing. We shall choose a slightly more difficult example, so that the reader can follow the process in full. Assume that the incident beam V_i is a partially and right-elliptically

polarized beam specified by the Stokes vector $\begin{bmatrix} 6 \\ 3 \\ 2 \\ 1 \end{bmatrix}$ and that the

polarizer is an ideal linear polarizer with transmission axis at $-45°$; its matrix M (according to Table 9-1) is:

$$\frac{1}{2}\begin{bmatrix} 1 & 0 & -1 & 0 \\ 0 & 0 & 0 & 0 \\ -1 & 0 & 1 & 0 \\ 0 & 0 & 0 & 0 \end{bmatrix}, \text{ that is, } \begin{bmatrix} .5 & 0 & -.5 & 0 \\ 0 & 0 & 0 & 0 \\ -.5 & 0 & .5 & 0 \\ 0 & 0 & 0 & 0 \end{bmatrix}.$$

To find the vector of the emerging beam, the task is to perform the multiplication

$$[V_e] = [M][V_i] = \begin{bmatrix} .5 & 0 & -.5 & 0 \\ 0 & 0 & 0 & 0 \\ -.5 & 0 & .5 & 0 \\ 0 & 0 & 0 & 0 \end{bmatrix} \begin{bmatrix} 6 \\ 3 \\ 2 \\ 1 \end{bmatrix}.$$

The first step is to multiply all four numbers ("elements") of the vector by the respective four elements of the *top row* of the matrix, then add the four products. The result is

$$0.5(6) + 0(3) - 0.5(2) + 0(1),$$

and the sum is 2. This is the first (top) element of the vector V_e of the emerging beam.

Next, the same four elements of the vector are multiplied by the respective four elements of the *second* row of the matrix, giving $0(6) + 0(3) + 0(2) + 0(1)$; the sum is 0. The same is then done for the third row, and the sum obtained is -2. Finally, the same is done for the fourth row, and the sum obtained is 0.

These four sums are the four respective elements of V_e, which describes the emerging beam. That is,

$$[V_e] = \begin{bmatrix} 2 \\ 0 \\ -2 \\ 0 \end{bmatrix}.$$

This is the answer. Reference to Chapter 8 shows that it implies an emerging beam that has intensity 2 and is 100 percent linearly polarized at $-45°$.

Another example: Suppose the same partially elliptically polarized beam V_i considered above strikes a linear quarterwave plate (i.e., a 90° linear retarder) the fast axis of which is at 45°. What is the outcome? The matrix of such a retarder is found from Table 9-1 to be

$$[M] = \begin{bmatrix} 1 & 0 & 0 & 0 \\ 0 & 0 & 0 & -1 \\ 0 & 0 & 1 & 0 \\ 0 & 1 & 0 & 0 \end{bmatrix},$$

TABLE 9-1 *Examples of Matrices That Represent Ideal Homogeneous Polarizers and Retarders*

Device	Mueller Matrix	Jones Matrix
Linear polarizer with transmission axis horizontal ($\theta = 0°$)	$\frac{1}{2}\begin{bmatrix} 1 & 1 & 0 & 0 \\ 1 & 1 & 0 & 0 \\ 0 & 0 & 0 & 0 \\ 0 & 0 & 0 & 0 \end{bmatrix}$	$\begin{bmatrix} 1 & 0 \\ 0 & 0 \end{bmatrix}$
Same, with axis vertical ($\theta = 90°$)	$\frac{1}{2}\begin{bmatrix} 1 & -1 & 0 & 0 \\ -1 & 1 & 0 & 0 \\ 0 & 0 & 0 & 0 \\ 0 & 0 & 0 & 0 \end{bmatrix}$	$\begin{bmatrix} 0 & 0 \\ 0 & 1 \end{bmatrix}$
Same, with axis at 45°	$\frac{1}{2}\begin{bmatrix} 1 & 0 & 1 & 0 \\ 0 & 0 & 0 & 0 \\ 1 & 0 & 1 & 0 \\ 0 & 0 & 0 & 0 \end{bmatrix}$	$\frac{1}{2}\begin{bmatrix} 1 & 1 \\ 1 & 1 \end{bmatrix}$
Same, with axis at −45°	$\frac{1}{2}\begin{bmatrix} 1 & 0 & -1 & 0 \\ 0 & 0 & 0 & 0 \\ -1 & 0 & 1 & 0 \\ 0 & 0 & 0 & 0 \end{bmatrix}$	$\frac{1}{2}\begin{bmatrix} 1 & -1 \\ -1 & 1 \end{bmatrix}$
Right circular polarizer	$\frac{1}{2}\begin{bmatrix} 1 & 0 & 0 & 1 \\ 0 & 0 & 0 & 0 \\ 0 & 0 & 0 & 0 \\ 1 & 0 & 0 & 1 \end{bmatrix}$	$\frac{1}{2}\begin{bmatrix} 1 & -i \\ i & 1 \end{bmatrix}$
Left circular polarizer	$\frac{1}{2}\begin{bmatrix} 1 & 0 & 0 & -1 \\ 0 & 0 & 0 & 0 \\ 0 & 0 & 0 & 0 \\ -1 & 0 & 0 & 1 \end{bmatrix}$	$\frac{1}{2}\begin{bmatrix} 1 & i \\ -i & 1 \end{bmatrix}$
Linear quarterwave retarder with fast axis horizontal ($\rho = 0°$)	$\begin{bmatrix} 1 & 0 & 0 & 0 \\ 0 & 1 & 0 & 0 \\ 0 & 0 & 0 & 1 \\ 0 & 0 & -1 & 0 \end{bmatrix}$	$\begin{bmatrix} e^{i\pi/4} & 0 \\ 0 & e^{-i\pi/4} \end{bmatrix}$
Same, with axis vertical ($\rho = 90°$)	$\begin{bmatrix} 1 & 0 & 0 & 0 \\ 0 & 1 & 0 & 0 \\ 0 & 0 & 0 & -1 \\ 0 & 0 & 1 & 0 \end{bmatrix}$	$\begin{bmatrix} e^{-i\pi/4} & 0 \\ 0 & e^{i\pi/4} \end{bmatrix}$
Same, with axis at 45°	$\begin{bmatrix} 1 & 0 & 0 & 0 \\ 0 & 0 & 0 & -1 \\ 0 & 0 & 1 & 0 \\ 0 & 1 & 0 & 0 \end{bmatrix}$	$\frac{1}{\sqrt{2}}\begin{bmatrix} 1 & i \\ i & 1 \end{bmatrix}$
Same, with axis at −45°	$\begin{bmatrix} 1 & 0 & 0 & 0 \\ 0 & 0 & 0 & 1 \\ 0 & 0 & 1 & 0 \\ 0 & -1 & 0 & 0 \end{bmatrix}$	$\frac{1}{\sqrt{2}}\begin{bmatrix} 1 & -i \\ -i & 1 \end{bmatrix}$
Linear halfwave retarder with fast axis at 0° or 90°	$\begin{bmatrix} 1 & 0 & 0 & 0 \\ 0 & 1 & 0 & 0 \\ 0 & 0 & -1 & 0 \\ 0 & 0 & 0 & -1 \end{bmatrix}$	$\begin{bmatrix} 1 & 0 \\ 0 & -1 \end{bmatrix}$
Same, with axis at ±45°	$\begin{bmatrix} 1 & 0 & 0 & 0 \\ 0 & -1 & 0 & 0 \\ 0 & 0 & 1 & 0 \\ 0 & 0 & 0 & -1 \end{bmatrix}$	$\begin{bmatrix} 0 & 1 \\ 1 & 0 \end{bmatrix}$
Circular halfwave retarder, right or left	$\begin{bmatrix} 1 & 0 & 0 & 0 \\ 0 & -1 & 0 & 0 \\ 0 & 0 & -1 & 0 \\ 0 & 0 & 0 & 1 \end{bmatrix}$	$\begin{bmatrix} 0 & 1 \\ -1 & 0 \end{bmatrix}$

and accordingly the outcome is computed by performing the multiplication:

$$[V_e] = [M][V_i] = \begin{bmatrix} 1 & 0 & 0 & 0 \\ 0 & 0 & 0 & -1 \\ 0 & 0 & 1 & 0 \\ 0 & 1 & 0 & 0 \end{bmatrix} \begin{bmatrix} 6 \\ 3 \\ 2 \\ 1 \end{bmatrix}.$$

Multiplication of the four elements of the vector by the four respective elements of the top row of the matrix yields the result $1(6) + 0(3) + 0(2) + 0(1)$, or 6. Multiplications involving the second, third, and fourth rows yield the results -1, 2, and 3. Thus the vector V_e of the emerging beam is

$$[V_e] = \begin{bmatrix} 6 \\ -1 \\ 2 \\ 3 \end{bmatrix}.$$

In this example so many of the elements are zero that the multiplication can indeed be done in one's head.

The vectors V_i and V_e in this example are intriguingly similar. They differ only in the second and fourth elements; these have been interchanged, and one sign has been altered. What the retarder does, essentially, is to interchange these two elements and alter the sign of one. What a succinct statement this is! The Stokes vector method of describing the beam is so "full of brevity and orthogonality" that—even in a situation in which a partially elliptically polarized beam is involved—the function of the matrix of the retarder is merely to interchange two numbers and alter the sign of one.

A close examination of the matrix reveals the cause of the interchange. In the first and third rows, the number one appears in a position lying directly on the main diagonal of the matrix, and accordingly the first and third elements of the vector are entirely unaffected. The one of the second and fourth rows appears *not* on the main diagonal, but shifted two spaces right or left from it; this is why the second and fourth elements of the vector are shifted two places vertically.

A person familiar with matrices can often "see ahead of time" what a given matrix will do to a given vector. This is especially true of the matrix of a retarder: by examining such a matrix he can infer at a glance which elements of the vector will be interchanged and which will undergo a change of sign. He will come to realize what a concise and eloquent thing the matrix is.

The Mueller calculus is at its very best in problems involving a long series, or train, of polarizers and retarders. Here, too, the method is simply one of routine matrix multiplication. For example, if there are four polarizers or retarders A, B, C, and D and their respective matrices are called M_a, M_b, M_c, and M_d, the vector V_e of the emerging beam is found by performing the straightforward multiplication:

$$[M_d][M_c][M_b][M_a][V_i].$$

Since matrices do not commute, they must be written in the correct order. The order employed above is correct if the incident beam strikes A first, then B, then C, then D. That is, the matrix that stands next to the vector V_i must be the matrix of the device that the incident beam strikes first. The procedure is to multiply V_i by M_a; then multiply the result by M_b; then multiply this by M_c; and so on. The final result is, again, a vector that indicates the emerging beam's intensity, degree of polarization, and form of polarization.

Telescoping the Matrices: Another pleasant aspect of the Mueller calculus should be noted—a bit of elegance and simplicity perhaps unsurpassed in physics. When a beam strikes a train of polarizers and retarders, the train of matrices can be telescoped into *one* matrix. There is no train so complicated that it cannot be represented, functionally, by just one matrix.

Telescoping the series of matrices $[M_d][M_c][M_b][M_a]$ is accomplished by multiplying M_a by M_b, then multiplying the product (which itself is a 4×4 matrix) by M_c, and then multiplying this result by M_d. The resulting matrix, called M_t, is *an accurate and complete description of all the capabilities of the train with respect to any incident beam* (of the given wavelength). No matter what the polarization form of the beam, this one matrix permits immediate, one-step evaluation of the over-all effect of the entire train! When many different incident beams are to be considered, much time is saved by using just the one telescoped matrix M_t each time. Of course, the sequence of the polarizers and retarders must not be altered, and the azimuths, too, must not be changed.

Sometimes the result obtained from telescoping a train of matrices is surprising. This applies, for example, to a train consisting of two linear, 180° retarders the fast axes of which are at 0° and 45°. The telescoping process proceeds thus, symbolically:

$$[M_b][M_a] = [M_t]$$

and thus in detail (using the appropriate matrices from Table 9-1):

$$\begin{bmatrix} 1 & 0 & 0 & 0 \\ 0 & -1 & 0 & 0 \\ 0 & 0 & 1 & 0 \\ 0 & 0 & 0 & -1 \end{bmatrix} \begin{bmatrix} 1 & 0 & 0 & 0 \\ 0 & 1 & 0 & 0 \\ 0 & 0 & -1 & 0 \\ 0 & 0 & 0 & -1 \end{bmatrix} = \begin{bmatrix} 1 & 0 & 0 & 0 \\ 0 & -1 & 0 & 0 \\ 0 & 0 & -1 & 0 \\ 0 & 0 & 0 & 1 \end{bmatrix}.$$

The last matrix is the telescoped matrix of the train. That is, functionally it represents the pair of 180° retarders in the given sequence and with the given orientations. The surprise is that it is identical to the matrix of a single 180° circular retarder. The train of two linear retarders acts like one circular retarder. This remains true, of course, irrespective of the polarization form of the incident beam. This example illustrates strikingly the insight that the matrix methods provide.

How are the Mueller matrices derived? One might guess that they stem from the electromagnetic theory, but this is not the case. They rest on a phenomenological foundation, i.e., on experiment. Professor Mueller has declared, only partly jokingly, that if his matrices should be found to disagree with the electromagnetic theory, ". . . so much the worse for the electromagnetic theory." Indeed, certain problems (involving scattering of polarized or partially polarized light) cannot be handled adequately by means of the electromagnetic theory, yet *can* be handled successfully with the aid of his empirically-arrived-at matrices.

The key phenomenon on which the Mueller matrices rest is the *linear* relationship that is found to exist between incident and emergent beams when there is, say, a polarizer or a retarder situated in the beam. Experiment shows that, under all normal circumstances, each property (*I*, *M*, *C*, or *S*) of the emerging beam depends on the first power of each property of the incident beam. Consequently, one can write a set of *linear* equations relating the properties of the emerging beam to the properties of the incident beam. Four constants appear in each equation; hence there are 16 constants in all. Because of the linear relationship mentioned, the same 16 constants apply regardless of the polarization form of the incident beam. Thus the set of 16 constants can be regarded as *the* specification of the given polarizer or retarder at the given orientation. *This set is the matrix.*

In some experiments performed with lasers, the intensity is so great that the relationships are no longer linear; under these circumstances matrices do not apply.

JONES CALCULUS

The Jones calculus has so much in common with the Mueller calculus that no detailed discussion is needed. Again, each polarizer and retarder—in its given orientation—is specified by a matrix. And again, the effect of a train of polarizers and retarders on a given beam is computed by multiplying the vector of the incident beam by the matrices of the polarizers and retarders. As before, the matrices must be written down in the correct order.

The vector that applies here is not the four-element Stokes vector, but the two-element Jones vector, described in Chapter 8. The matrices used are not the 4×4 Mueller matrices, but the 2×2 Jones matrices. Some commonly used, normalized, Jones matrices are shown in Table 9-1. Many of these are extremely simple; for example, the matrix of an ideal horizontal linear polarizer is $\begin{bmatrix} 1 & 0 \\ 0 & 0 \end{bmatrix}$, the matrix of a right-circular polarizer is $\begin{bmatrix} 1/2 & -i/2 \\ i/2 & 1/2 \end{bmatrix}$, and the matrix of a 180° linear retarder with the fast axis horizontal is $\begin{bmatrix} 1 & 0 \\ 0 & -1 \end{bmatrix}$.

The Jones calculus has several unique advantages. One is the avoidance of redundancy. The general Jones matrix contains four elements comprising, in all, eight numbers, and none of these numbers is a functon of any other. The Mueller matrix, on the other hand, contains much redundancy: it includes 16 numbers, and only seven of these are independent (seven rather than eight, because no information as to absolute phase is included).

Also, every normalized Jones matrix that can be written down corresponds to a device that can be produced in the laboratory. Thus, because there are eight numbers in the general matrix, eight distinct classes of optical devices can be specified—and indeed eight such classes can be produced in the laboratory. These exhibit:

(1) Linear retardance, fast axis horizontal;
(2) Linear retardance, fast axis at 45°;
(3) Circular retardance, fast eigenvector clockwise;
(4) Linear polarizance, axis horizontal;
(5) Linear polarizance, axis at 45°;
(6) Circular polarizance, major eigenvector clockwise;
(7) Isotropic phase change; and
(8) Isotropic absorption.

The most general Jones matrix implies a device exhibiting all eight of these properties (or the opposite properties) to some extents. The same is *not* true of the Mueller matrix: to write down a Mueller matrix that corresponds to nothing known in the laboratory is all too easy. An example is the matrix

$$\begin{bmatrix} 1 & 0 & 0 & 0 \\ 0 & 0 & 0 & 0 \\ 0 & 0 & 0 & 0 \\ 0 & 0 & 0 & 0 \end{bmatrix}$$; the first element is unity and hence implies a rea-

sonably large transmittance for any incident, polarized beam; yet all the other elements are zero, implying that the emerging beam is entirely unpolarized. In fact no device of this general type, i.e., no ideal depolarizer, exists.

Another unique feature of the Jones matrix is that it can be differentiated, to yield information as to the intensive properties of the material comprising the polarizer or retarder in question. The processes of squaring, cubing, etc., or even integrating are sometimes both possible and useful.

However, the Jones calculus is applicable only if the incident beam is already polarized and only if the polarizers and retarders are of nondepolarizing type. Scattering devices cannot be handled by this calculus.

The Mueller calculus and the Jones calculus are so new that they are mentioned in few textbooks. However, they are coming into use rapidly, and they appear to have a bright future. Their brevity and compactness appeal to theoretical physicists, and their easy tabulation and manipulation appeal to applied physicists.

10　*Popular Applications of Polarized Light*

If there is a logical order in which the various applications of polarizers and polarized light should be considered, the authors have never discovered it. The policy adopted here is to consider the most popular and "humanistic" applications first, and the more scientific and esoteric applications last.

POLARIZATION AND THE HUMAN EYE

The most humanistic fact about polarization of light is that it can be detected directly by the naked eye. Nearly anyone, if told carefully what to look for, can succeed in this. Sometimes he can even determine the form and azimuth of polarization.

What the observer actually "sees" is a certain faint pattern known as Haidinger's brush and illustrated in Fig. 10-1. The brush is so faint and ill-defined that it will escape notice unless the field of view is highly uniform: a clear blue sky makes an ideal background, and a brightly illuminated sheet of white paper is nearly as good. The best procedure for a beginner is to hold a linear polarizer in front of his eye, stare fixedly through it toward a clear blue sky, and, after five or ten seconds, suddenly turn the polarizer through 90°. Immediately the brush is seen. It fades away in two or three seconds, but reappears if the polarizer is again turned through 90°. The brush itself is sym-

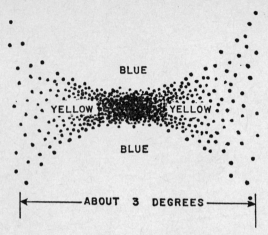

FIG. 10-1 Approximate appearance of Haidinger's brush when the vibration direction of the beam is vertical.

metric, double-ended, and yellow in color; it is small, subtending an angle of only about 2° or 3°. The adjacent areas appear blue, perhaps merely by contrast. The long axis of the brush is approximately perpendicular to the direction of electric vibration in the linearly polarized beam, i.e., perpendicular to the transmission axis of the polarizer used.

Circular polarization, too, can be detected directly by eye, and even the handedness can be determined. When an observer facing a clear blue sky places a right circular polarizer in front of his eye, he sees the yellow brush and finds that its long axis has an upward-to-the-right, downward-to-the-left direction, i.e., an azimuth of about +45°. This is true, of course, irrespective of the orientation of the polarizer, since a circle has no top or bottom. If he employs a *left* circular polarizer, he finds the brush to have a −45° orientation. In each case the pattern fades away rapidly, but can be restored to full vigor by switching to a polarizer of opposite handedness. Instead of using a circular polarizer the observer can use a single linear polarizer in series with a 90° retarder, the latter being held nearer to the eye. Turning the retarder through 90° reverses the handedness of the circular polarization.

Some people see the brush easily; others have difficulty. A few see the brush when looking innocently at the partially polarized blue sky, i.e., without using any polarizer at all, and even without meaning to see the brush. Some people see the brush more distinctly by linearly polarized light than by circularly polarized light, and for others the reverse is true. An observer may find the brush to have a slightly different orientation depending on which eye is used.

The spectral energy distribution of the light is important. If the light is rich in short-wavelength (blue) radiation, the brush is very noticeable, but if the short-wavelength radiation is eliminated by means of a yellow filter, the brush fails to appear. Use of a blue filter tends to accentuate the brush.

Although the phenomenon was discovered in 1844, by the Austrian mineralogist Haidinger, the cause is not yet fully understood. Presumably the thousands of tiny blue-light-absorbing bodies in the central (foveal) portion of the retina are dichroic and are oriented in a radial pattern, for example, a pattern such that the absorption axis of each body lies approximately along a radius from the center of the fovea. Incident linearly polarized light will then be absorbed more strongly in some parts of the pattern than in other parts and consequently some parts will fatigue more than others. When the vibration direction of the light is suddenly changed, the varying degrees of fatigue are revealed as a subjective radial pattern. Presumably no such dichroism or orientation pattern applies to longer wavelength (yellow and red) light; consequently a yellow sensation dominates in those regions where fatigue-to-blue has occurred.

The fact that circular polarization, also, may be detected perhaps implies that some transparent portion of the eye is weakly birefringent and acts like a retarder, converting circularly polarized light to linearly or elliptically polarized light. The direction of the major axis of the ellipse depends only on the direction of the fast axis of the retarding layer and hence remains fixed—unless the observer tips his head.

Perhaps physicists will some day write matrices to describe the retarding layers and dichroic areas of the eye. Poets were the first to see magic fire and jewels in the human eye; physicists will be the first to see matrices!

NAVIGATION BY BEES

Bees, too, can detect the vibration direction of linearly polarized light. The experiments of the biologist K. von Frisch during World War II showed that bees "navigate" back and forth between hive and source of honey by using the sun as a guide. More interesting, when the sun is obscured by a large area of clouds the bees can still navigate successfully if they can see a bit of blue sky: they can detect the azimuth of linear polarization of the blue light and navigate with respect to it. One way of demonstrating the bee's ability to detect the azimuth of polarization is to place the bee in a large box the top of which consists of a huge sheet of linear polarizer, such as H-sheet. Each time the experimenter turns the polarizer to a different azimuth, the bee changes his direction of attempted travel correspondingly.

Certain other animals also can detect the polarization of sky-light and navigate by it. This includes ants, beetles, and the fruit fly *Drosophila*. Probably many other examples will be discovered.

POLARIZATION OF SKY LIGHT

Blue-sky light traveling in a direction roughly at right angles to the sun's rays is partially polarized. When an observer holds a linear polarizer in front of his eye and gazes in a direction perpendicular to the direction of the sun, he finds that rotating the polarizer slowly causes the sky to change from bright to dark successively. The degree of polarization of sky light may reach 70 or 80 percent when the air is clear and dust-free, the sun is moderately low in the sky, and the observation direction is near the zenith.

The polarization is a result of the scattering of the sun's rays by the molecules in the air. Rayleigh's well-known inverse-fourth-power law relating scattering intensity to wavelength accounts for the blue color of the scattered light, and the asymmetry associated with the 90° viewing angle accounts for the polarization, as explained in Chapter 5. Some multiple scattering occurs, and this reduces the degree of polarization somewhat; when the observer ascends to a higher altitude, the amount of air involved is reduced, multiple scattering is reduced, and the degree of polarization is increased. A further increase results when a yellow or red filter is used to block the short-wavelength component of the light and transmit the long-wavelength component —the latter component is less subject to multiple scattering. (The situation is very different for infrared radiation of wavelength exceeding 2 microns: much of this radiation is produced by emission from the air itself, rather than by scattering, and this exhibits little or no polarization.)

Some persons are capable of detecting the polarization of sky light directly by eye, by virtue of the Haidinger brush phenomenon discussed in a preceding section; a few individuals find the brush noticeable enough to be a nuisance. Ordinarily, of course, it escapes notice and plays little part in the affairs of man. Its practical use by bees, ants, etc., has been indicated, and the importance to photographers is discussed in a later section.

POLARIZATION OF LIGHT UNDER WATER

A surprising fact about the polarization found in light present beneath the surface of the ocean (or of a pond) is that the predominant direction of electric vibration is horizontal. The opposite might be expected, since most of the light that enters the water enters *obliquely* from above, and the most strongly reflected component of obliquely incident light is the horizontally vibrating component. But oceanographers and biologists, working at depths of 5 to 30 feet in waters off Bermuda and in the Mediterranean Sea, have found the main cause of submarine polarization to be the scattering of the light by microscopic particles suspended in the water. Sunlight and sky light enter the water from above, and the average direction of illumination is roughly vertical; consequently the polarization form of the scattered light that travels horizontally toward an underwater observer is partially polarized with the electric vibration direction horizontal. The situation is much the same as that discussed in Chapter 5, except that the incident light has a more steeply downward direction and the asymmetric scattering is by microscopic particles instead of molecules.

Typically, the degree of polarization is 5 to 30 percent, an amount found to be important to a variety of underwater life. The water flea *Daphnia* tends to swim in a direction perpendicular to the electric vibration direction, for reasons not yet known. When tests are conducted in a tank filled with water that is free of suspended particles, so that the submarine illumination is practically unpolarized, *Daphnia* ceases to favor any one direction. But if suspended matter is added, thus restoring the polarization, *Daphnia* resumes the custom of traveling perpendicular to the vibration direction.

The arthropod *Limulus* (horse shoe crab) easily detects the polarization of the underwater light and is presumed to navigate with respect to the electric vibration direction. The same is true of the crustacean *Mysidium gracile* and various other forms of marine life. Most tend to swim perpendicularly to the vibration direction; some swim parallel to it; a few swim at different relative orientations depending on the time of day. For all of these animals, polarization is a compass that works even under water!

POLARIZING SUNGLASSES

The lenses of ordinary sunglasses employ absorbing materials that are isotropic, and accordingly the incident light is attenuated by a fixed factor irrespective of polarization form. This is unfortunate. The fact is that "glare" consists predominantly of light having a horizontal vibration direction. Why? For these reasons:

(a) The main source of light (sun and sky) is overhead, and consequently the main flux of light is downward.

(b) The surfaces that are most strongly illuminated by the downward flux are horizontal surfaces.

(c) Such surfaces are usually viewed obliquely, since a person seldom looks straight down.

(d) Most outdoor objects are of dielectric material.

(e) Light reflected obliquely from a horizontal dielectric surface is partially linearly polarized with the dominant vibration direction *horizontal,* as explained in Chapter 4.

Polarizing sunglasses take full advantage of this fact. The lenses are made of dichroic material (H-sheet, usually) oriented with the transmission axis vertical, as indicated in Fig. 10-2a, so

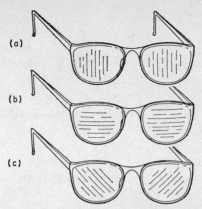

FIG. 10-2 Three types of polarizing spectacles. In (a) the transmission axis is vertical, for eliminating glare reflected from horizontal surfaces. In (b) the axis is horizontal, for eliminating reflections from vertical windows of trains, store-fronts (show-windows), etc. In (c) the axis directions are 45° and −45°, a standard arrangement used in viewing polarization-coded stereoscopic pictures.

that almost all of the horizontal vibrations are absorbed. The component having vertical vibration direction is transmitted. Usually some isotropic absorber is included in the lenses to absorb ultraviolet light strongly and blue and red light to a moderate extent; the sunglasses then have a greenish hue which has nothing to do with the polarization.

Motorists and vacationists find that polarizing sunglasses are helpful not only in reducing the brightness of the field of view as a whole, but also in enhancing the beauty of the scene. Because specularly reflected light is absorbed preferentially, roads, trees, grassy fields, etc., appear softer and more deeply colored through polarizers. Specularly reflected light tends to veil nature's inherent beauty; polarizing sunglasses remove the veil.

Fishermen and boatsmen enjoy another benefit from wearing polarizing sunglasses. They want to be able to see fish, rocks, etc., beneath the surface of the water, yet the light from such objects is dim and is usually lost in the "noise" of the sky light reflected obliquely from the surface. Since the reflected light is highly polarized with horizontal vibration direction, the polarizing sunglasses absorb this component strongly, and the visibility of the underwater objects is greatly increased. The increase is greatest when the viewing direction corresponds to the polarizing angle, which, for water, is about 53° from the normal. When the viewing direction is along the normal, i.e., straight down, there is no increase at all.

There is one interesting situation in which polarizing sunglasses produce little increase in visibility of underwater objects even when the angle of viewing is the polarizing angle. This situation occurs when the sky is clear and blue, the sun is low in the sky, and the pertinent portion of the sky is at 90° from the direction of the sun. Under these circumstances the light striking the water is already linearly polarized at such an azimuth that almost none of it is reflected. There is no task left for the sunglasses to perform—there is no reflected glare to suppress. The underwater objects are seen with great clarity. Persons unfamiliar with the polarization of sky light and with the dependence of oblique reflection on polarization form are likely to ascribe the remarkable clarity to "especially clear water" rather than to absence of reflection.

CAMERA FILTERS

Photographers often wish to enhance the contrast between blue sky and white clouds. Thirty years ago they did this by employing a yellow filter, which absorbed most of the blue light from the clear sky but transmitted most of the white light from the clouds. Using ordinary black-and-white film, they obtained excellent contrast by this method. Today, photographers are using color film increasingly, and the use of yellow filters is no longer permissible since it would eliminate all blue colors from the finished photograph.

The only known solution is to exploit the difference in polarization between blue sky and white clouds. Light from most portions of the blue sky is partially linearly polarized, as explained in a preceding section, and light from clouds is unpolarized. Therefore a neutral-color, linear polarizer mounted at the optimum azimuth in front of the lens will absorb a large fraction (e.g., 80 percent) of the sky light while transmitting a large fraction (nearly half) of the light from the clouds; thus the contrast is increased by a factor of two or three. The factor is less if the air is hazy, and more if the air is extremely clear (as in Arizona) and if the camera is aimed about 90° from the direction of the sun.

The usual way of choosing the azimuth of the polarizer is crude, but perhaps adequate. The photographer holds the polarizer in front of his eye, finds by trial and error which azimuth maximizes the contrast of the clouds in question, and then attempts to mount the polarizer on the camera without changing the azimuth of the polarizer. One type of polarizing filter for cameras is equipped with a small "satellite" polarizer mounted at the end of a short arm and aligned permanently with the main polarizer. The photographer installs the main polarizer in front of the lens, looks through the small polarizer and turns the arm to whatever azimuth maximizes the contrast. Both polarizers then have this optimum orientation. The satisfactoriness of the azimuth can be checked visually at any time. Instead of using these empirical methods, a scientifically minded photographer can proceed by dead reckoning, i.e., by following this well-known rule: Mount the polarizer so that its transmission axis lies in the plane determined by camera, sun, and object photographed. (So oriented, the polarizer performs a valuable additional service: it eliminates most of the specularly reflected light from trees, roads, etc., and enhances the softness and depth of color of the scene.)

When a photographer standing on a sidewalk tries to photograph objects situated behind a store window, the reflection of the street scene from the window may threaten to spoil the photograph. An excellent solution is to place the camera off to one side so that the window is seen obliquely at about the polarizing angle, and mount a linear polarizer in front of the lens; the polarizer is turned so that its transmission axis is horizontal, and the polarized light reflected from the window is absorbed. The authors have a friend who has applied this same principle to a pair of special spectacles he wears while touring the country by railroad. The lenses consist of polarizers oriented with the transmission axis horizontal, as indicated in Fig. 10-2b; thus when he gazes out of the train window in oblique forward direction, the reflected images of passengers and newspapers are wiped out, and the scenery appears in its pristine glory.

USE OF CIRCULAR POLARIZERS IN ELIMINATING PERPENDICULARLY REFLECTED LIGHT

Eliminating perpendicularly reflected lights is a different problem from that of eliminating obliquely reflected light. The process of oblique reflection at Brewster's angle causes the reflected beam to be linearly polarized, and accordingly a linear polarizer can eliminate the reflected beam entirely. But the process of *normal* reflection, i.e., with incident and reflected beams *perpendicular* to the smooth glossy surface in question, produces no polarization at all. How, then, can the specularly reflected light be eliminated while light originating behind the surface is transmitted freely?

The question is an important one to radar operators scanning the cathode-ray-oscilloscope screens on which dim greenish spots representing airborne objects appear. The screen proper is situated in a large evacuated tube, and the greenish light emerges through a curved glass window at the front end of the tube. (Sometimes the window is flat; sometimes a safety plate of glass or plastic is mounted close in front of it.) Often the operator has difficulty in seeing the greenish spots, not only because they are faint, but also because they may be masked by various extraneous images reflected by the front surface of the window, e.g., reflections of room lights and of people, clothing, papers, etc., situated near the operator. Extinguishing the room lights would eliminate these reflections, but would make it impossible for the operator to read instructions or make notes. What he needs is some kind of filter that will transmit the light originating behind the window and absorb the light reflected approximately perpendicularly from it.

This need is filled by the circular polarizer. Such a device, if mounted close in front of the window, will transmit nearly half of the light that originates behind the window, yet will eliminate about 99 percent of the room light that is reflected perpendicularly from it. The circular polarizer acts on the room light *twice:* it circularly polarizes room light that is approaching the window, then absorbs the reflected component. The logic behind this requires explanation. Two key facts must be kept in mind:

(1) A beam that is reflected perpendicularly and specularly by a smooth glossy surface has the same degree of polarization as the incident beam, since the reflection process does not introduce randomness of any kind.

(2) The reflection process reverses the handedness of polarization, because handedness is defined with respect to the propagation direction and the reflection process reverses the propagation direction.

If the polarizer is of right-circular type, as in the arrangement shown in Fig. 10-3, room light that passes through and ap-

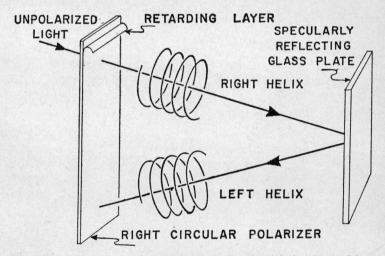

FIG. 10-3 Use of a circular polarizer in absorbing light reflected by a surface approximately perpendicular to the incident beam. Note that the reflection process reverses the handedness of circular polarization.

proaches the window is right-circularly polarized; the reflected light is *left*-circularly polarized and hence is *totally absorbed by the polarizer*. In effect, the circular polarizer "codes" the light, the window reverses the coding, and the polarizer then annihilates the reverse-coded beam. If both faces of the window are ideally flat and smooth, if the light is incident exactly along the normal, and if the polarizer is truly of circular type, the reflected light is totally absorbed. Usually the conditions are less ideal: the rear surface of the window usually serves as support for the luminescent screen and has a matte appearance; the window is usually curved and much of the troublesome room light incident on the window makes an angle of 10° or 20° or more with the normal; and the polarizer, although circular with respect to some wavelengths, is elliptical with respect to others. Nevertheless, the improvement provided by the polarizer is large, and the amount of faint detail that the operator can see on the screen is greatly increased.

One precaution must be mentioned: reflections from the polarizer itself must be avoided. This is usually accomplished by tilting the polarizer forward so that the only reflected images the observer sees are images of a dark-colored floor or other dark objects.

Television sets, also, have been equipped successfully with circular polarizers. If the set is used in a brightly lit room, or is used outdoors, the circular polarizer performs a valuable service in trapping the specularly reflected glare and thus increasing the picture-vs-glare ratio by a factor of the order of 10.

VARIABLE-DENSITY FILTER

A pair of linear polarizers arranged in series is an almost ideal device for controlling the transmitted intensity of light. Rotating one polarizer through an angle θ with respect to the other causes the intensity of the transmitted light to vary approximately as $\cos^2 \theta$. Because the transmittance is easily varied and easily calculated, the pair of polarizers has found much favor in the eyes of designers of spectrophotometers and other devices for controlling and measuring light intensity.

Specially designed sunglasses employing pairs of linear polarizers in place of lenses have been used successfully by aviators and others. One polarizer of each pair can be rotated through an angle as large as 90°, and a linkage connecting the two pairs insures that the attenuation is the same for both eyes. By moving one small lever, the wearer can vary the transmittance throughout a range of about 10,000 to 1.

Controllable pairs of very-large-diameter polarizers have been used as windows of railroad cars and ocean liners. A person sitting near such a window turns a small knob to rotate one polarizer with respect to the other and thus reduce the intensity of the transmitted light to any extent desired.

One of the authors has experimented with a variable-density filter employing *three* linear polarizers in series, in order that a transmittance range of 10^8 to 1 could be achieved. The device worked well and, as expected, obeyed a cosine-fourth, rather than a cosine-square law.

THREE-DIMENSIONAL PHOTOGRAPHY AND THE USE OF POLARIZERS FOR CODING

Millions of polarizers found their way into the motion picture theaters of North America in 1952 and 1953 when stereoscopic (three-dimensional, or 3-D) movies achieved brief prominence. Each spectator wore a pair of polarizing spectacles called viewers, and polarizers were mounted in front of the projectors.

A photographer who enjoys looking at 3-D still pictures in his living room needs no polarizers. Usually he employs a small viewing box containing a light source and two lenses, one for each eye; a black partition, or septum, divides the box into right and left halves. The picture, consisting of two small transparencies mounted about two inches apart in a side-by-side arrangement on a cardboard frame, is inserted in the box so that the right-eye transparency lines up with the right lens and the left-eye transparency lines up with the left lens. (The two transparencies are, of course, slightly different because they were taken by cameras situated about two or three inches apart; the spacing used approximates the spacing of the two eyes.) The side-by-side arrangement of the two transparencies and the presence of the septum insure that the observer's right eye sees only the right transparency and the left eye sees only the left transparency. No cross-communication, or "cross-talk," can occur. Consequently the observer enjoys an impressively realistic stereoscopic illusion.

When 3-D motion-picture films are projected in a theater, many complications arise. Separate projectors must be provided for the right-eye and left-eye movie films, and the two projectors must be synchronized within about 0.01 second. Since there is just one large screen and this is to be viewed by hundreds of spectators, there can be no septum. Indeed, no practical geometrical method of preventing cross-talk is known.

Before the advent of mass-produced polarizers in the 1930's, an *analglyph* system of preventing cross-talk was invented. It applied wavelength coding to the two projected beams. The right-eye picture was projected through a long-wavelength (red) filter, and the left-eye picture was projected through a shorter-wavelength (green) filter. The spectator's viewers contained right and left lenses of red and green plastic, respectively, and accordingly each lens transmitted light from the appropriate projector and absorbed light from the other. Thus each eye received just the light intended for it. The system succeeded as a short-term novelty: stereoscopic illusions were created. But the system had two major defects: chromatic "retinal rivalry" between the two eyes, and incompatibility with the showing of colored motion pictures.

In the 1930's the problem was solved with éclat by a polarization-coding system, demonstrated with great impact at the New York World's Fair of 1939 and improved in later years. As indicated in Fig. 10-4, a linear polarizer oriented with its transmission axis at $-45°$ is placed in front of the projector used for the right-eye pictures, and a polarizer at $+45°$ is placed in front of the projector used for the left-eye pictures. Thus the two beams striking the movie screen are orthogonally coded. The lenses of the spectator's viewers consist of correspondingly oriented linear polarizers, and so each eye receives only light that originates in the appropriate projector. Superb stereoscopic illusions result. Since the polarizers perform well at all wavelengths in the visual range, color movies can be presented as easily and faithfully as can black-and-white movies.

The polarizers placed in front of the projectors consist, ordinarily, of K-sheet; as explained in Chapter 3, K-sheet is highly resistant to heat, and any polarizing filter placed close in front of a powerful projector is bound to heat up considerably since it necessarily absorbs about half the light. The lenses of the 3-D viewers are usually of HN-38 sheet; it has high major transmittance k_1 and small minor transmittance k_2, and it is inex-

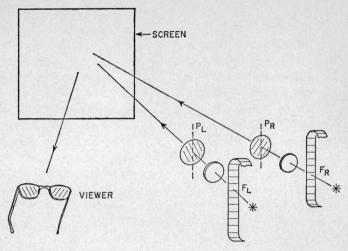

FIG. 10-4 Arrangement for projecting polarization-coded stereoscopic motion-picture films by means of two side-by-side projectors. Films F_R and F_L containing the "right-eye pictures" and "left-eye pictures" are mounted in the right and left projectors, which are equipped with linear polarizers P_R and P_L oriented at $-45°$ and $+45°$ respectively. The viewer contains correspondingly oriented polarizers, and accordingly each eye sees only the images intended for it.

pensive. The viewers are cheap enough (about 10¢ each) that they can be discarded after a single use.

The polarization-coding scheme has one limitation: if the spectator tilts his head to one side, the polarizers in his viewers no longer line up accurately with the respective polarizers on the projectors. Thus cross-talk occurs: the right eye sees faintly the image meant for the left eye, and vice versa: each eye sees a faint ghost image in addition to the main image. The spectator does not enjoy this. The difficulty could be avoided if the linear polarizers were replaced by high-quality, achromatic circular polarizers, but unfortunately no method is known for producing achromatic circular polarizers economically.

The effectiveness of any polarization-coding projection system is destroyed if the screen depolarizes the light appreciably. Screens that have a smooth aluminum coating usually conserve polarization to the extent of about 99 percent, but those having a matte white surface or a rough metallic coating produce much depolarization and hence much cross-talk between the two images. Many of the screens used in the innocent days of 1952 and 1953 were of the wrong type, and the resulting ghost images were a major annoyance. For that reason, and because of frequent lack of care in maintaining synchronism between the two projectors, movie-goers soon turned back to conventional 2-D pictures. Some nostalgia remains, however. Persons who were lucky enough to see a full-color, 3-D movie showing attractive actors filmed against a background of gorgeous scenery look forward to the time when well-made, well-presented 3-D movies, with their almost miraculous realism and intimacy, will animate the theaters once again.

THE VECTOGRAPH

The type of three-dimensional photography discussed in the preceding section is parallel-projected 3-D photography. The two motion-picture films are situated side-by-side, and two projectors are operated in parallel. During the late 1930's a radically new approach, called *vectography*, was developed by E. H. Land, J. Mahler, and others. In this system, the two films are arranged in series, bonded together. Because of the permanent series arrangement, many problems disappear. Only one projector is needed, and perfect synchronism is "guaranteed at the factory." Each pair of pictures (each vectograph) is projected as a single unit, in the same projector aperture and at the same time, and onto the same area of the same screen. If the film breaks, it can be spliced with no concern as to preservation of synchronism.

The method can succeed only if means are provided for preserving the identity of the two coincident projected beams. Again, polarization-coding is the answer. However, because the two images are bonded together in series, the coding must occur within the images themselves. In the system used by Land and Mahler each image consists of varying quantities of linearly dichroic molecules aligned in a common direction, and the directions employed in the two images are mutually at right angles. Dark areas in any one image contain a high concentration of dichroic molecules; light areas contain little or no dichroic material; but irrespective of concentration, the alignment direction is always the same. For the other image, the alignment direction is always orthogonal to the first. It is to be noted that the images contain no silver and no other isotropic absorber. Only aligned absorbers having high dichroic ratio are used.

A communications engineer would describe the vectograph by saying that it provides two distinct channels. Each is assigned to one image. Each is independent of the other. Since the vectograph images themselves perform the polarization coding, no polarizer is used in front of the projector; indeed, the interposition of such a filter would play havoc with the system. As before, the screen must preserve the polarization and the spectator's viewers must perform the appropriate decoding, or discriminating, act. Excellent stereoscopic effects are achieved. However, the production of vectograph film is a costly undertaking involving very specialized equipment, and constant attention is needed to maintain high enough dichroic ratio so that the channels are truly independent and ghost images are avoided.

Vectograph pictures of the "still" type are easier and cheaper to make than vectograph movies. Stereo pairs of aerial photographs of mountainous country, if presented in vectograph form, give a navigator (wearing an appropriate viewer) a very realistic impression of the terrain, and a map maker can prepare an accurate contour map from the vectograph with ease.

POLARIZING HEADLIGHTS

It is ironic that the main goal of Land and others in developing high-quality, large-area, low-cost polarizers has never been achieved. The polarizers are used with great success in dozens of applications, but not the application that was uppermost in the minds of the inventors.

Their goal was to eliminate glare from automobile headlights. In an era when dual-lane highways, circumferential bypasses, and other safety engineering advances were virtually unknown and the aim and focus of automobile headlights were highly erratic, the glare that confronted motorists at night was almost unbearable, and was an important cause of accidents. As early as 1920 several illumination engineers recognized that the glare could be eliminated by means of polarizers—if large-area polarizers could somehow be produced. If every headlight lens were covered by a linear polarizer oriented with the transmission axis horizontal and every windshield were covered with a linear polarizer oriented with its axis vertical, no direct light from the headlights of Car A could pass through the windshield of oncoming Car B. Drivers in both cars could see road-markings, pedestrians, and so forth, but neither would experience any glare from the other's headlights. Moreover, it would be permissible for each driver to use his *high* beam continuously, and accordingly his ability to see pedestrians, etc., would be greater than before, despite the fact that each polarizer would transmit only about half of the light incident on it.

It was soon recognized that the analyzing polarizer should not be made a permanent part of the windshield, but should be incorporated in a small visor situated just in front of the driver's eyes. During the day, when headlights were not in use, the visor could be swung out of the way. It was also recognized that care should be taken to make sure the headlight polarizers had sufficient light-leak, i.e., sufficiently large k_2 value, that the headlights of oncoming cars would not disappear entirely!

Land and his colleagues moved rapidly. They invented a whole series of polarizers, each superior to its predecessor. The first successful type, J-sheet, employed aligned, microscopic crystals of the dichroic mineral herapathite; the method of manufacture is described in Chapter 3. Then came H-sheet, which was better in nearly every respect and in addition was easier to make. Finally, K-sheet appeared; it had most of the superb qualities of the earlier materials and the added virtue of being unaffected by fairly high temperature, such as 215°F. To persons seeking polarizers for use in headlights, K-sheet appeared to be the pot of gold at the end of a polarized rainbow.

Concurrently, several better ways of orienting the polarizers were proposed. One attractive scheme was to orient the headlight polarizers and the visor polarizer at the identical azimuth, namely —45°, as indicated in Fig. 10-5. Then, even a polarization-conserving object in the path of the headlights would appear

FIG. 10-5 Automobile equipped with headlight polarizers and a visor polarizer oriented at −45°. When two such cars approach one another, each driver is protected from the glare from the headlights of the other.

to the driver to be brightly illuminated. (This would not be the case if his visor polarizer were crossed with his headlight polarizers.) The −45° system disposed of the headlight glare problem adequately: if two cars A and B both equipped in this manner approached one another at night, each driver's visor would be crossed with the other car's headlight polarizers, and neither driver would experience any glare.

Using the Mueller calculus, Billings and Land compared a wide variety of polarizer orientation schemes, and found several to be particularly attractive. Perhaps the best system was one called "−55°, −35°." The transmission axes of the headlight polarizers and visor polarizer are at 55° and 35° from the vertical, respectively, an arrangement that minimizes complications stemming from the obliquity of the portion of the windshield situated just in front of the driver.

Despite the successes on all technical fronts, the project bogged down. To this day no one knows just why. Probably many little reasons were responsible. Among these were the following:

(1) The polarizers absorbed slightly more than half of the light incident on them, and accordingly the automobile manufacturers felt that they would have to increase the power of the lamps themselves and perhaps use larger generators and batteries also.

(2) Some windshields were moderately birefringent; therefore

they would act like retarders, alter the polarization form of the incident light, and allow some glare to leak through.

(3) Nearly every year the automobile manufacturers increased the backward tilt of the windshields; such tilt tends to alter the polarization form of light having an oblique vibration direction, and hence leads to glare-leak.

(4) Passengers, as well as drivers, would require visors, since passengers also dislike glare.

(5) Pedestrians might find that the glare was worse than ever, unless they too employed polarizing visors or spectacles.

(6) The system would succeed only if adopted by *all* car manufacturers, and therefore no one manufacturer would gain any promotional advantage from it.

(7) The first few drivers to put the system to use would get little benefit from it for at least a year or two, i.e., until millions of other cars were similarly equipped.

(8) It was difficult to decide when and how to force the owners of old cars to install the necessary polarizers on their cars.

(9) The patents on the only fully satisfactory polarizers were held by a single company.

(10) To introduce the system would require formal, coordinated action by all States.

(11) Improvements in headlight design and aiming, the increasing numbers of dual-lane highways, and the brighter street lamps used in cities and suburbs led some people to believe that the need for a polarization-type of glare control was no longer acute.

However, persons who have actually experienced the polarization method of glare removal are convinced that the drawbacks are trivial compared to the benefits.

Perhaps some day the system will be tried out on a pilot scale in a small, isolated community, where all the cars could be equipped with polarizers in a few weeks. Perhaps an island of moderate size would make a good test ground. If the system is found to be highly successful there, it will presumably spread throughout every country that teems with automobiles.

11 *Scientific and Esoteric Applications*

POLARIMETRY AND OPTICAL ACTIVITY

In 1811 the French scientist Arago discovered that when linearly polarized light travels through certain materials the vibration direction changes, but the type of polarization remains unchanged. Today many such materials are known. A familiar example is an aqueous solution of the common sugar *dextrose*. As the linearly polarized light travels along in the solution the vibration direction shifts progressively in clockwise, or right-handed sense as judged by an observer looking through the solution toward the light source. For example, the vibration direction might be 0° at the location where the light enters the solution, 10° one inch farther along, and 20° at a point another inch farther yet. The important fact is that the light remains linearly polarized; no conversion to elliptical or circular form occurs.

The shift, or *rotation angle* ζ, of the vibration direction is found to be proportional to the concentration C of sugar in the solution and to the pathlength L. That is,

$$\zeta = CL[\alpha].$$

The proportionality constant $[\alpha]$ is called the *rotatory power* of the solution. Since it is a function of wavelength, monochromatic light must be used when the rotation angle is to be measured accurately.

This formula is highly prized by the sugar industry. If a chemist wishes to know the concentration of sugar in a given solution, he pours some of the liquid into a glass cell, passes a beam of linearly polarized light through it, and measures the extent ζ to which the vibration direction is altered. Using the formula, he can at once compute the concentration.

The apparatus used in measuring optical rotation is called a *polarimeter*. Many types of polarimeter have been invented. The simplest employs two linear polarizers mounted at either end of the glass cell. One polarizer is fixed and the other can be rotated in its own plane. Before filling the cell, the operator orients the latter polarizer at whatever angle θ_1 causes the light passing through the cell to be extinguished; that is, he "crosses" it with the fixed polarizer. He then pours the solution into the cell and turns the adjustable polarizer to whatever angle θ_2 re-establishes extinction. Then $\zeta = \theta_2 - \theta_1$. The method is success-ful, but not very accurate: the operator has trouble deciding when the extinction is most nearly perfect. His precision is of the order of 0.1°.

In a high-precision polarimeter, a *split-field polarizer* is used. This consists of a disk the two halves of which are linear polar-izers that have been oriented at slightly different angles. Figure 11-1a shows a design in which the transmission axes of the two

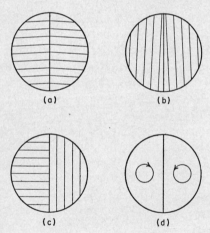

(a)   (b)

(c)   (d)

FIG. 11-1 Split-field polarizers: (a) 3°-3° polarizer; (b) 87°-87° polar-izer; (c) 0°-90° polarizer; (d) left-right polarizer, or circular dichroscope. The transmission-axis directions of the linear polarizers are indicated by the hatch marks.

side-by-side polarizers make angles of 6° with one another and angles of ±87° with the dividing line between them. The device is mounted in fixed manner, with the dividing line vertical, at the far end of the glass cell. The operator looks through the rotatable polarizer at the near end of the cell and observes the split-field polarizer at the far end. On finding that one half of the field appears darker than the other, he turns the adjustable polarizer until both halves appear equally dark, and records its angle. Now the precision of setting is very high, thanks to the fact that the two halves of the field are compared simultaneously and the disappearance of the dividing line between them is a sensitive criterion of equality of brightness.

Polarimeter designers have spent much time trying to find the optimum difference in axis direction of the two halves of the split-field polarizer. Should the directions differ by 5°, 10°, 20°, or some other angle? There is now agreement that the answer depends on the quality of the polarizers and the brightness of the light source: if the polarizers have high polarizance and the light source is bright, 5° may be an excellent choice. But if the polarizers are leaky and the light source is dim, a 10° or 20° angle may be superior. Instead of making a "butt joint" between the two polarizers, the designer may start with a single disk-shaped polarizer and then cement to one half of it a small polarizer set at a slightly different angle. Sometimes a small prism-type polarizer, called a Lippich prism, is used for this purpose.

The split-field polarizer is scarcely an exciting device, but it stands as a landmark among simple but precise optical tools.

In recent years the precision achievable in polarimetry has been increased by the substitution of a photocell for the human eye. A single photocell can scan the two halves of the field sequentially and repeatedly, or two photocells may be used to compare the two halves simultaneously. A precision of 0.001° can be achieved. When photocells are used, ultraviolet or infrared radiation may be used in place of visible light; also, the dependence of rotation angle on wavelength can be explored throughout a wide spectral range.

As stated in Chapter 6, materials that rotate the vibration direction are called rotators and are said to exhibit optical activity. How is the rotation accomplished? The question is easily answered. Optical activity is an old friend in new clothing: it is merely circular birefringence. Any circularly birefringent body divides an incident, linearly polarized beam into right and left circularly polarized components and retards one relative to the other. Thus when the two components are combined as they emerge from the body, the resultant, although still a straight line, has a different orientation. If the two refractive indices and the pathlength are known, the retardance is easily computed by means of the formula presented in Chapter 6. The rotation angle ζ is simply half the retardance, $\delta/2$. For example, if the right-circularly polarized component is retarded by $45°$ relative to the left-circularly polarized component, the azimuth of the resultant is found by averaging the phases $-45°$ and $0°$ of the two components; that is, $\zeta = (0° - 45°)/2 = -22\frac{1}{2}°$.

Many kinds of liquid solutions are optically active. Since a liquid has no axes, the amount of rotation per unit pathlength in a given solution is the same no matter what the direction of the incident beam. In other words, the optical activity is inherent in the individual molecule of the dissolved material (the solute). The activity demonstrates that the individual molecule has a *handedness*. Chemists trying to find the structure of new and complicated molecules take full advantage of this; many substances of interest to organic chemists in general and biochemists in particular have a handedness.

Many crystals, too, are optically active. In most instances the activity is due to the arrangement of the atoms in the crystal lattice. The amount of optical rotation depends strongly on the propagation direction in the beam, and if the crystal is dissolved the optical activity disappears. A quartz crystal exhibits optical activity provided the light is traveling approximately parallel to the optic axis; actually there are two kinds of quartz crystals, one of which rotates the light to the right and the other to the left. The crystal cinnabar is famous for the large amount of rotation it produces, namely, about $5800°$ per centimeter of pathlength. Certain so-called liquid crystals, e.g., cholesteryl benzoate, produce a rotation exceeding $250\,000°$ per centimeter.

Most gases do *not* ordinarily exhibit optical activity. However, they become active when subjected to a strong magnetic field. This subject is discussed in the following section.

THE FARADAY EFFECT

In 1845 the English physicist Michael Faraday discovered that certain substances that ordinarily are optically *in*active become active when situated in a strong magnetic field. The rotation is largest when the field is parallel to the light beam and vanishes when the field is perpendicular to the beam. His discovery created much interest since it provided the first direct link between light and electromagnetism. Faraday predicted that the link would "prove exceedingly fertile," and Maxwell's work a generation later bore this out.

A unique feature of the magnetically induced rotation, or Faraday effect, is that if the light is reflected directly backwards, so as to retrace its path (in the region of magnetic field) the rotation does *not* cancel. The opposite is true of a sugar solution: if the vibration direction of a beam is rotated 20° when the light is traveling east in a sugar solution it is rotated 20° in the opposite direction when reflected west; the over-all or net rotation is zero. But a rotation that depends on an applied magnetic field is not reversible, and indeed if the light is reflected back and forth repeatedly the rotations are additive, and the over-all rotation increases without limit. Physicists trying to measure the Faraday rotation produced by a gas take advantage of this fact; they use concave mirrors to reflect the light back and forth many times through the region of magnetic field. If the combined path-length amounts to at least 10 meters and if the strength of the magnetic field amounts to several thousand gauss, measurable rotation is produced by almost every known gas.

The amount of the Faraday rotation depends strongly on the frequency of the radiation. The rotation of ultraviolet radiation is relatively large, and that of infrared radiation and radio waves is small. Nevertheless, radioastronomers find the effect bothersome: they find that linearly polarized radio waves arriving from distant stars have passed through so many billions of miles of space in which traces of gas and weak magnetic fields exist that the vibration direction has been altered drastically. This complicates the task of finding what the *original* vibration direction was.

PHOTOELASTIC ANALYSIS

Photoelastic analysis is the technology of transforming patterns of mechanical strain into patterns of light. It is a translation from the world of the invisible to the world of the visible. The layman cares little about strains, and if some small utensil in his home breaks from excessive strain, he mutters and buys another. But to the mechanical engineer who designs the utensil, the locations and magnitudes of the strains are important. If he can measure the strains under typical conditions of "loading," he can find how to change the design so as to reduce the likelihood of breakage. In designing simple girders and bridges, engineers can calculate the strains by means of formulas listed in engineering handbooks. But many objects of commercial importance have complicated shapes, and the calculations are too difficult. Here photoelastic analysis comes into its own. It is so successful that at least 10 full-length books have been written on this subject alone.

The physical phenomenon that makes photoelastic analysis possible is the creation of birefringence in a transparent object when the object is strained. A square glass rod, for example, becomes birefringent when a tensile force is applied to each end. It becomes, in effect, a linear retarder. To interpret the retardation in terms of strain, the engineer invokes these elementary facts:

(1) The two axes of the retarder indicate directly the two principal axes of strain, i.e., the directions parallel and perpendicular to the strain.

(2) The magnitude of the retardance is proportional to the magnitude of the strain.

Thanks to the phenomenon of *strain-birefringence,* attributable to the distortions of the electrical structures of the molecules, the task of evaluating strains deep within the rod is reduced to that of evaluating a retardation pattern that can be examined outside the rod.

The method is applicable not only to rods, but to transparent objects of almost any size and shape. To find the strains in an *opaque* object, e.g., a steel gear, an engineer resorts to a transparent model, or stand-in: he makes a similar gear of glass or plastic and conducts the test on it. The models are usually made of plastic rather than glass, since plastics are more easily machined and, in addition, provide larger retardance for a given strain. The plastic pieces used for this purpose must have no "built-in" birefringence such as sometimes results from improper annealing in manufacture.

To *see* the retardation pattern, the engineer places the object in a *polariscope.* This usually consists of little more than a crossed pair of linear polarizers consisting, perhaps, of two huge sheets of HN-32 polarizer. A large, bright light source is placed behind the polariscope. The two polarizers are mounted an ample distance apart, and the test object is placed between them. If the object is a glass disk, and if no force is applied to it, an engineer looking through the assembly finds the entire field of view to be dark; the glass is isotropic, hence has no retardance, and accordingly the crossed polarizers extinguish the light. But when a steady compressive force is applied at each end of a diameter of the disk, various regions of the disk become strained to different extents and in different directions. Each region acts as a linear retarder and "defeats the extinction" to an extent that depends on the axis-directions of the retarder and on the amount of retardance. Thus the engineer sees a large pattern of light areas and dark areas, or *fringes.*

From a single glance at the optical pattern an engineer can often tell a lot about the strain pattern. He can tell where the strain is greatest and can estimate the approximate magnitude. With a little effort he can determine the direction and magnitude of the strain at each point with considerable accuracy. He recognizes that any region that appears black irrespective of how the disk is turned is a retardance-free region and hence is a region of no strain at all (except perhaps a directionless hydrostatic strain). A region S that appears bright when the disk is at certain orientations clearly has considerable retardance; the fast and slow axes (hence also the principal axes of strain) may be determined merely by finding the two 90°-apart orientations that produce extinction; when the disk is at either of these orientations, the strain axes of region S are parallel to the axes of the crossed polarizers.

The exact magnitude of the retardance of region S can be

found accurately by several methods. One is to introduce various calibrated retarders and find which just cancels the retardance in question. Another method is to notice the color of the light transmitted by S; since the retardance varies with wavelength, different magnitudes of strain produce different colors (as well as different intensities); elaborate correlations between retardance and color have been discovered and tabulated. A third method, applicable when the retardance amounts to many cycles, entails "counting the fringes" between region S and some region known to be strain-free.

The final step, that of converting retardance values to strain values, is made with the aid of proportionality constants called *stress-optic constants*. The constants applicable to various common kinds of glass and plastic are listed in textbooks on photoelastic analysis. Of course, the retardance values must take into account the thickness of the transparent body.

Interesting variations of the method have been devised. Crossed right- and left-circular polarizers may be used instead of crossed linear polarizers; then, all regions that have the same strain are observed to have the same brightness irrespective of the direction of strain. Another variation, applicable when the test object is very thick and of complicated shape, is to "freeze the strain in," then cut the object into slices and study each separately. Another method, useful if the object is of steel or other opaque material and is so complicated that to construct a model of plastic would be impractical, is to apply a mirror-like coating to the object, then apply a firmly bonded plastic film of known thickness, and then view the object (by means of, say, a circular polarizer) by reflected light. When the steel object is strained, the plastic film is strained also, and by studying the retardation pattern of the film the engineer can derive much information as to the strain in the underlying opaque object.

Photoelastic analysis can be applied even to high-speed events such as the impact of a bullet on a plastic plate. The time delay between application of the mechanical force and onset of the retardance pattern is so small (less than a microsecond) that the correlation between strain and retardance remains valid. If the retardation pattern is recorded by high-speed photography, the strains can be evaluated from the photographs.

LYOT FILTER

The Lyot filter was invented in 1932 to fill a long-felt need of astrophysicists. Until then, they had no adequate way of photographing the solar corona or solar prominences by monochromatic light. Using combinations of glass filters of different colors, they could restrict the bandwidth of the light reaching the photographic film to about 300 Å, but this was far too broad an interval. Spectroheliographs provided a bandwidth of about 1 Å, but were unsatisfactory in many ways. What was needed was a simple, passive filter having a 1 or 2 Å passband centered at, say, the prominent 6563 Å line Hα of the Balmer series of hydrogen. If several such filters were available, each providing a passband at a different wavelength, astrophysicists could obtain a series of solar photographs indicative of the concentrations and distributions of several different elements. Such information would open the door to solving problems relating to temperature, energy transport, etc.

The problem was solved by a 25-year-old French astronomer, B. F. Lyot. Abandoning hope of finding materials having sufficiently selective absorption bands, he found a way of exploiting *chromatism of birefringence* with dramatic success. He designed a train of retarders and polarizers that had high transmission for light of wanted wavelength (6563 Å) and almost no transmission for other wavelengths. Each retarder was situated between a pair of polarizers. Being of chromatic type, the retarders provided spectral discrimination in terms of ellipticity and azimuth of vibration; absorption of light of nonprivileged vibration direction was accomplished by the polarizers. Lyot used bulky, prism-type polarizers, since sheet polarizers had not yet become available, but in 1938 Y. Öhman of Sweden exhibited a more compact design employing sheet polarizers. We shall describe the latter design.

As indicated in Fig. 11-2, the Lyot-Öhman filter includes six linear polarizers and five linear retarders. The polarizers are oriented with their transmission axes at 45°; the retarders, consisting of plates of quartz cut parallel to the optic axis, are oriented with their fast axes horizontal. The important fact is that each of the series of retarders is twice as thick as its predeces-

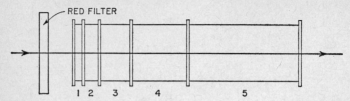

FIG. 11-2 The Lyot-Öhman filter. The five retarders of progressively greater thickness are separated by linear polarizers. Light passes through the assembly from left to right.

sor, and hence has twice the retardance and *twice the variation of retardance with wavelength.* The significance of this design becomes clear when the operation of the first (thinnest) retarder is understood. This element receives light linearly polarized at 45° and transforms it to varying extents, depending on the wavelength; the thickness of the retarder is such that the retardance with respect to light of the wanted wavelength (6563 Å) is exactly 360°—i.e., one complete cycle—and consequently this light is transmitted freely by the next polarizer, with transmission axis parallel to that of the first. Other wavelengths, however, are retarded to various different extents, emerge polarized with various azimuths and ellipticities, and are absorbed by the following polarizer to various extents; certain wavelengths are absorbed entirely. The uppermost curve of Fig. 11-3 shows that the spectral transmittance curve is approximately sinusoidal; in the spectral range included in the graph there are two broad peaks and two broad valleys. The second retarder is twice as thick and has a retardance of 720° for light of the wanted wavelength, which emerges unchanged in vibration direction and passes freely through the next polarizer while light of unwanted wavelengths suffers various changes in polarization form and is absorbed to varying extents; the transmission curve has four peaks and four valleys. The third retarder is twice as thick again, and has a retardance of 1440° for the wanted wavelength, which again is transmitted freely by the following polarizer while other wavelengths are absorbed to various extents. And similarly for the succeeding retarders—each discriminates more sharply against wavelengths close to the wanted wavelength. (For simplicity of

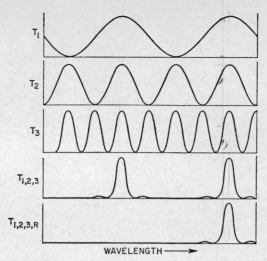

FIG. 11-3 Spectral transmission curves of a simplified, three-retarder Lyot-Öhman filter. Curves T_1, T_2, T_3 pertain to the three segments individually; $T_{1,2,3}$ pertains to the combination of the three; $T_{1,2,3,R}$ is the curve that results when the red filter is included.

exposition, only three retarders are included in the filter of Fig. 11-3.)

The combined effect of all the retarders and polarizers is spectacular: almost the entire spectrum is wiped out. Only two slender bands remain. One of these is now eliminated by means of a conventional filter of appropriately colored glass, and *voila!* —a single, slender transmission band remains. Typically its width is 1 to 3 Å.

The filter achieved wide acclaim and remained supreme for many years. However, a rival design was put forward in 1953 by the Czechoslovakian scientist I. Šolc: it includes only two polarizers and, between them, a stack of perhaps 60 identical chromatic retarders each with its fast axis shifted relative to that of the preceding layer. Again a remarkably narrow passband is achieved. Which design is superior? Careful analyses (employing the Jones calculus) and comparisons have been made; the conclusion is that the merits are almost equal, but the Lyot-Öhman design is slightly superior.

In recent years great advances have been made in the art of fabricating narrow-band filters of the interference type, i.e., filters employing spectral selectivity of reflection, and the Lyot-Öhman filter has suffered partial eclipse. However, in 1961 three Australian physicists, W. H. Steel, R. N. Smartt, and R. G. Giovanelli, gave this filter a new lease on life by improving the design to the extent that a passband as narrow as 0.13 Å is achieved. The incident beam must be collimated, and the temperature of the filter must be regulated accurately. The filter includes eight polarizers, of KN-40 sheet, and no less than 48 retarders of quartz and calcite.

MEASURING THE TORQUE PRODUCED BY CIRCULARLY POLARIZED LIGHT

In 1935 R. A. Beth, working at Princeton University, tackled a job that physicists had been avoiding for decades. Long before 1900 physicists had realized that circularly polarized light must exert a torque on any body that absorbs the light. Using the wave theory of light they had computed, in 1899, the amount of torque to be expected. When the quantum theory came along a generation later, new computations were made, with the same general outcome. Physicists spoke confidently of the spin and angular momentum of the circularly polarized photon, but until 1935 no one had ever measured its torque.

Beth was aware of the difficulties. The *pressure* exerted by a beam of circularly polarized light on a small test body is impressively small; but the *torque* is much smaller still. If the torque were to be detected at all, the test body would have to be suspended from the end of a quartz fiber so long and thin that the natural period of torsional oscillation would be of the order of ten minutes. Mechanical vibrations, air currents, temperature variations, and stray electrical fields could easily mask whatever twist might be produced by the circularly polarized light. To avoid these disturbances Beth employed a massive vibration-free stand, mounted the delicate equipment in an almost perfect vacuum, and surrounded it with a thick copper shield.

To maximize the torque produced by the circularly polarized light, he used two ingenious tricks. The first was to employ, as test body, a 180° linear retarder, rather than an absorbing body. An absorbing body would merely absorb the light and respond straightforwardly to the torque. But the retarder transmitted the light and reversed the handedness; because the light was transmitted, unwanted heating effects were avoided, and because the handedness was reversed, the torque was doubled. The second trick was to place a mirror just beyond the retarder, to reflect the light back through it and double the torque again.

A powerful projection lamp was used, and also a condensing lens. The light beam, aimed vertically upward toward the 180° retarder, was linearly polarized by a Nicol prism, and a 90° linear retarder mounted at 45° or 135° was used to convert the polarization form to right- or left-circular.

Beth's elaborate preparations paid off. When he illuminated the retarder with right- and left-circularly polarized light alternately, at the resonant frequency of the fiber suspension system, the retarder began to oscillate with steadily increasing amplitude. When he used unpolarized light or linearly polarized light, no oscillation was produced. Thus for the first time the torque produced by circularly polarized light was demonstrated. The magnitude of the torque was found to be about 10^{-9} dyne-centimeter, a value in close accord with predictions. Physicists' confidence in their understanding of circular polarization was greatly strengthened.

PRODUCING A GIGANTIC SHIFT IN THE WAVELENGTH OF POLARIZED LIGHT

Anyone can change the polarization form of a polarized light beam without changing the wavelength: all he has to do is insert a retarder in the beam. But can he change the wavelength without changing the polarization? Obviously he can make a trivial change in wavelength easily enough: he can reflect the light from a moving mirror, or he can move the source rapidly toward or away from the observer. But can he produce a *large* change in wavelength without changing the polarization form?

High-energy physicists in several laboratories believe the answer is yes, and are preparing experiments to prove they are right. R. H. Milburn and others from Tufts University are pre-

paring to perform such an experiment at the nearby Cambridge
Electron Accelerator Laboratory. Their goal is to produce a
beam of polarized photons each having an energy of about 800
million electron volts (800 MeV). The proposed method is to
start with a beam of linearly polarized *visual-range* photons and
then *increase their energy by a factor of about 4 × 10⁸ while
leaving the polarization undisturbed.* If their efforts succeed, the
beam of polarized 800-MeV photons will be put to use by physicists trying to discover the fundamental nature of matter and
energy. At present, polarized very-high-energy photon beams are
unknown. If such beams are produced, they will give a great boost
to high-energy physics.

Milburn's scheme for producing a beam of monochromatic,
polarized, 800-MeV photons is fantastic in the extreme. It consists of taking a beam of monochromatic, vertically polarized,
visual-range photons and aiming it head-on at a beam of 6-BeV
electrons traveling in the opposite direction at almost the speed
of light, specifically, $0.999\,999\,99\ c$. Some of the visual-range
photons should collide squarely with the electrons and recoil with
greatly increased energy, *but without losing their polarization.*
Persons familiar with the Compton effect will recall that when a
photon collides with a stationary electron an exchange of energy
occurs: the electron gains energy and the photon loses energy.
But if the electron is initially traveling at tremendous speed the
outcome is very different: the electron loses energy and the photon gains energy. In the proposed experiment the gain is almost
unbelievably large: the energy increases from about 2 eV to 800
MeV, a 400-million-fold increase. The collision of electron and
photon may be likened to a collision between a fast truck and a
fast ping-pong ball: the lighter particle recoils with greatly increased energy. (There is the paradox that the photon, although
recoiling with vastly increased energy, is traveling no faster than
before the collision.)

How can the polarization of the photon remain unchanged during a collision in which the energy is increased so drastically?
Will not the polarization be "lost in the shuffle"? Apparently
not. If the light is regarded as a wave train with vertical vibration direction, the effect of the train is to make the individual
electron oscillate up and down, so that the radiation it emits

has the same (vertical) vibration direction. The Doppler shift is enormous, and explains the 400-million-fold increase in frequency and in photon energy.

The experiment has its Achilles' heel, namely the small probability that any one photon will collide with an electron. However, it is estimated that, if a high-intensity laser is used as the light source, if the beam is nicely aligned with the oppositely traveling electrons, and if the pulse of photons from the laser is timed to coincide with the pulse of 10^{11} high-energy electrons from the accelerator, perhaps 10^3 collisions will occur each second. If this proves to be the case, the marriage of very-low-energy physics to very-high-energy physics will have a bright future.

MISCELLANEOUS APPLICATIONS

The number of scientific applications of polarizers and polarized light is enormous. There are far too many applications to describe even in a book several times the length of this one. A few of the outstanding applications that have *not* been discussed above are listed below.

Kerr cell: a very-fast-acting shutter consisting of an electrically controlled retarder situated between two crossed polarizers. Capable of opening and closing within 10^{-7} second, such an electro-optic shutter is used in high-speed photography and in modulating light beams from lasers.

Polarizing microscope: a microscope that is equipped with polarizers above and below the specimen and hence is capable of revealing the retardation pattern of a transparent specimen. Living cells and microscopic crystals of many kinds are so highly transparent as to be practically invisible when examined under an ordinary microscope, but many of them are birefringent and display much interesting detail when viewed under a polarizing microscope.

Analyzer of streaming birefringence: a device for measuring the retardance of a streaming liquid containing long, thin, birefringent molecules that have been aligned by the streaming process. From the gross retardance of the liquid a chemist can compute the birefringence of the individual molecule.

Analyzer of decay-time of fluorescence: a device for measuring
the delay involved in the emission of fluorescent light by a sub-
stance being irradiated with short-wavelength, linearly polarized
light. The longer the delay, the lower is the degree of polarization of the fluorescent light. From his measurements of the degree of polarization, the investigator can compute the decay
time, which may be as short as 10^{-9} second.

Illumination system for directing the growth of plants: By controlling the direction of electric vibration in the light illumi-
nating certain growing organisms, biologists can control the direction in which growth occurs. This applies, for example, to the
spores of fungi and the fertilized egg cells of algae.

Equipment for mapping the magnetic fields on the sun's disk:
By scanning across the sun's disk with a telescope equipped with
right and left circular polarizers and a photocell, astrophysicists
can measure the extent of Zeeman splitting of a spectral line
emitted by iron atoms and thus determine the intensity of magnetic field at any desired location on the disk. The method is
sufficiently rapid and precise that the changes occurring in the
magnetic field pattern from day to day can be followed.

*Equipment for evaluating the directions of magnetic fields in
distant nebulae:* By photographing distant nebulae with the aid
of linear polarizers, astronomers have discovered that in some
instances the light is partially linearly polarized. Almost 100
percent polarization is found in the light from many portions
of the Crab nebula. The pattern of magnetic field can be in-
ferred from the polarization pattern. The degree of polarization
of the light is so large that astrophysicists are convinced that portions of the nebula act like gigantic synchrotrons, accelerating
charged particles to speeds close to that of light and, as a by-
product, producing large amounts of polarized electromagnetic
radiation in the visual, infrared, and radio ranges. (The mechanism of emission of polarized light by a synchrotron is discussed in the following chapter.)

12   *Direct Generation of Polarized Light*

LOOK, MA, NO POLARIZER!

It is possible to produce polarized light directly, without benefit of polarizers. Some of the methods are new and extraordinary, and are discussed in the following sections. Others, older and duller, are discussed at the end of the chapter.

THE LASER

Perhaps the most impressive man-made device for the direct generation of polarized light is the laser, a type of light source unknown 10 years ago, but now enjoying widespread fame because of the high intensity produced and the highly monochromatic and highly coherent nature of its radiation. The laser is extraordinary in many respects. It is a cold lamp: high temperature is not required. It is a nonelectric lamp: no wires are connected to the lamp proper. (This is not true of certain more recent types of lasers.) It is a "chain-reaction" lamp: unless conditions are favorable, no chain reaction occurs and the lamp does not lase. It is an "auto-simplifying" lamp: as the chain reaction grows, the emitted beam becomes simpler and simpler, that is, it becomes more perfectly collimated, more monochromatic, more coherent, and *more highly polarized*. It is a long, thin, ungainly lamp, and when viewed from the side it scarcely appears to be lit at all. Yet it produces a beam intense enough to melt a hole in a razor blade in 1/1000 of a second. If this beam were to strike the eye of an observer, it would burn a hole in the retina equally rapidly.

Figure 12-1 shows the design of a gaseous laser intended to produce an intense monochromatic beam of linearly polarized light.

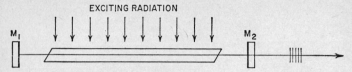

FIG. 12-1 Laser employing a gas-filled tube and two mirrors M_1 and M_2. The emitted light, emerging via partially transmitting mirror M_2, is linearly polarized with vibration direction in the plane of the paper. The atoms within the tube are excited by shorter-wavelength radiation incident transversely.

The heart of the device is a gas-filled tube with a mirror at each end. The type and the state of gas are crucial: before an intense beam having the desired photon energy E_1 can be produced, billions of atoms in the gas must be in an identical, metastable, excited state of energy E_1, achieved by means indicated in a later paragraph. The chain reaction, or *photon-breeding process*, may be started by causing a few photons having just this energy E_1 to travel through the gas. They trigger some of the excited atoms, inducing them to release their energy as additional photons of energy E_1. Some of these travel parallel to the axis of the tube, emerge from the tube, strike a mirror, and are reflected back along the tube again, triggering additional atoms.

Now the auto-simplifying process comes into play. The two mirrors have previously been oriented so as to reflect an axial ray back and forth through the gas many times, and they have been spaced so that the optical pathlength between them corresponds to an integral number of wavelengths, preferably a wavelength that is centrally located with respect to the very small but real range of variability in E_1. Accordingly, the radiation traveling parallel to the axis of the tube constitutes a standing wave pattern which dominates the breeding process and evokes from the excited atoms vast additional quantities of photons having the same direction, wavelength, and phase. In other words, the resulting amplification exhibits strong preference for the dominant direction, dominant wavelength, and dominant phase of the standing wave system. It is *coherent amplification* par excellence. Photons having slightly different specifications make fewer passes through the gas, produce fewer additional photons, and play an ever-diminishing role. This is especially true if the number of excited atoms is close to marginal, so that families of photons having less favorable specifications are starved.

How to extract a fraction of the energy from the standing wave pattern is a problem. It is usually solved by allowing one of the mirrors to transmit a few percent of the light incident on it. Eventually, cumulatively, a large fraction of the energy in the standing wave pattern passes through this mirror and is available for whatever use the experimenter has in mind.

It is the obliquity of the windows at the ends of the tube that insures that a single polarization form will dominate the chain reaction. The obliquity angle is chosen to be the polarizing angle, defined in Chapter 4. Consequently, a selection process comes into play as the light travels back and forth along the laser, repeatedly encountering the oblique windows. The component that has an electric vibration direction *in the plane of the diagram* passes through the window without loss and is amplified strongly in successive trips along the tube. But the component having the orthogonal vibration direction suffers reflection losses and becomes increasingly insignificant.

In some lasers a square-ended solid rod of transparent crystal of very special type is employed, rather than a gas. Even these *solid state* lasers will produce highly polarized light if the crystal is birefringent and has been cut so that the optic axis is perpendicular to the rod. Since the optical pathlength between mirrors is different for the two refractive indices, two different families of standing wave patterns are involved. The radiation in any one mode is usually polarized; the polarization form is sometimes linear, sometimes elliptical. In general, any laser that includes a birefringent rod or birefringent end-coating, or that employs asymmetry of reflection, is likely to produce light that is polarized to some extent.

To produce a sufficiently large population of atoms in a single, metastable, excited state is perhaps the central problem in laser design. Many ingenious solutions have been found, but will not be discussed here. Most commonly, excitation is accomplished by flooding the tube or rod transversely with shorter-wavelength radiation that the ground-level atoms absorb strongly; the atoms then find themselves in one or more unstable states with energy much greater than E_1, and many of the atoms promptly radiate enough energy so that they are left in the desired metastable state with energy E_1. Much skill is required in choosing a suitable kind of atom, identifying the various states concerned, and designing optical systems that will pour in enough of the shorter-wavelength radiation. In many lasers the population of excited atoms decreases drastically within the first few microseconds of the pulse, and operation must be halted for a second or so to allow time for the population to be built up again. Sometimes the size of the population oscillates rapidly and the output intensity oscillates also, exhibiting "intensity spikes."

The variety of lasers being manufactured is increasing rapidly, and an experimenter wishing to obtain a highly collimated, highly coherent, highly monochromatic, *highly polarized* beam of light finds the available range of devices to be increasingly attractive.

ELECTRON SYNCHROTRON

In terms of intensity the laser is the outstanding generator of polarized light. But in terms of time-averaged power, the electron synchrotron ranks first. No other man-made device can match its output of polarized short-wavelength radiation, which embraces the visual, ultraviolet, and x-ray spectral ranges.

An electron synchrotron is a circular machine for accelerating electrons to extremely high speed for use as "bullets" in high-energy physics experiments on fundamental particles. The largest such machine, the 6-BeV Cambridge Electron Accelerator at Cambridge, Mass., accelerates electrons to a speed of 99.999 999 percent the speed of light. They travel in a 240-ft-diameter horizontal orbit defined by a vast ring of electromagnets, one of which is pictured in Fig. 12-2. As the electrons gather speed (under the influence of high-intensity electric fields) their momentum and mass increase, and the strength of the vertical magnetic field is increased correspondingly in order to force the electrons to remain in the prescribed orbit. Magnetic focusing is employed to compress the cross section of the orbiting beam to about that of the lead in a lead pencil.

Each of the orbiting electrons is constantly being accelerated radially inward. Because it is accelerated, it radiates—as required by electromagnetic theory. When the speed of the electron

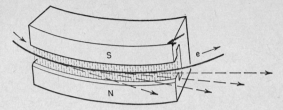

FIG. 12-2   Simplified diagram of one of the 48 magnets of the Cambridge electron synchrotron. The magnetic field is vertical and makes the path of the electrons curve to form part of a circular orbit. The polarized synchrotron radiation is emitted as a tangential "spray," indicated by the dashed arrows.

is moderate, the radiation, called *synchrotron radiation,* is emitted in all directions that are roughly perpendicular to the acceleration direction, i.e., to the radius vector. The radiation pattern, or lobe, has the "pumpkin shape" shown in Fig. 12-3a. No radiation is emitted in the directions of the cusps representing the ends of the pumpkin since these lie along the radius vector. But when the electron speed begins to approach the speed of light, relativistic effects distort the pattern enormously: it becomes stretched out, or bent around, in the forward tangential direction, and the two cusps also become displaced in the forward direction as indicated in Fig. 12-3b. Actually, the pattern, and the distortion, are slightly different for different wavelengths in the emitted radiation. When the electron speed reaches 99.999 999 percent of c, the pattern becomes almost unrecognizable. The pumpkin has been deformed into such a slender cone that, as illustrated in Fig. 12-3c, the cone would resemble a single line unless its width were exaggerated by the draftsman: the actual half-angle of the cone is less than $0.1°$. Cusps that define (for any one wavelength) directions of no-energy-propagation still exist, but are compressed within a small region of the cone; they now resemble two nostrils and are too slender to be detected. In summary, the radiation from any one small segment of the orbit is emitted practically along one single line, and is more highly collimated than the beam produced by a searchlight equipped with a large parabolic mirror; yet here no reflector is involved.

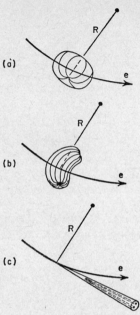

FIG. 12-3 Pattern, or lobe, of synchrotron radiation emitted by electrons in one small segment of the orbit. In (a) the electron speed is moderate, and the lobe is pumpkin-shaped, with the pumpkin ends, or cusps, lying along the radius vector R. In (b) the electron is traveling faster, and the lobe is distorted. In (c) the speed of the electron is close to that of light, and the lobe resembles a very slender cone; the two cusps are extremely close together.

It is remarkable that the electromagnetic theory, developed long before accelerators or relativistic electrons were conceived of, remains fully valid here. Starting with the basic laws of this theory, physicists have succeeded in deriving formulas that permit accurate prediction of the intensity, spectral energy distribution, and angular distribution of the radiation emitted by orbiting electrons traveling at speeds nearly equal to that of light. Theory and fact were compared closely by physicists operating the 300-MeV Cornell synchrotron; they found the agreement to be well within the limits of error of the experimental procedures. They are confident that the theory is valid also for the 6-BeV electrons in the Cambridge Electron Accelerator, although no attempt has yet been made to verify this. Theoretical calculations indicate that the wavelength range of the radiation from the 6-BeV electrons extends from about 0.1 Å to more than 100,000 Å, i.e., from the x-ray region into the infrared region, a total range of more than 20 octaves. Although the intensity of visible light is great and that of ultraviolet light is greater yet, the intensity of x-radiation is orders-of-magnitude greater still. The peak of intensity lies near 0.5 Å. The total output amounts to about *10 kilowatts of pure radiation;* it varies with the fourth power of the electron energy and thus sets an upper limit on the energy that is practical to achieve in an electron synchrotron of given diameter.

When examined at high resolution, the spectrum appears continuous. It is thus very different from the spectra of various common ultraviolet and x-ray sources. Theoretical physicists believe that, if all the electrons had identical energy, the spectrum would consist of millions of narrow bands, each a different harmonic of the orbital frequency of the electrons. But in practice the electron energies vary enough so that the banded nature of the spectrum is obscured.

When the light emitted exactly in the horizontal plane is examined carefully, it is found to be 100 percent linearly polarized, with the electric vibration direction horizontal. This finding is in agreement with electromagnetic theory. Actually, the polarization form can be simply predicted: because the acceleration is in the horizontal plane, the electric vector of the light must lie in this same plane. The experimental verification is complicated by three facts: the radiation originates in a slender "doughnut" of stainless steel and hence is not readily available for scrutiny; the synchrotron radiation is so intense that it promptly spoils the surface of any mirror or window inserted in the doughnut; and the entire neighborhood of the doughnut is pervaded by gamma radiation, neutron radiation, etc., intense enough to kill a man. Furthermore, because the operation of the synchrotron is cyclic, with a period of only 1/60 second, the properties of the radiation are continually changing, even from one millisecond to the next.

Although the light emitted in the horizontal plane is polarized linearly, the light emitted at about 0.25° above or below the horizontal plane is polarized elliptically. When the angle is increased slightly, the polarization form approaches circularity. It is a striking fact that the electrons emitting the light are *free* electrons; no atoms are involved.

In a synchrotron of small diameter the electrons travel around the orbit so fast and pass by a given observation window so many thousands of times in a given millisecond, that a real possibility exists that a dark-adapted observer could see the emitted light even when the population of the stream of electrons is reduced to *one single electron*. Literally he would then be seeing an individual electron, a feat that would overshadow even the historic Millikan oil-drop experiment. Already, physicists operating a small synchrotron at Frascati, Italy, have found that when there are very few electrons in orbit and a multiplier phototube is used to measure the intensity of the synchrotron radiation, the output of the photomultiplier decreases, from moment to moment, *in steps,* each step being of the same size and corresponding, apparently, to the loss of a single electron from the stream.

Persons who have been vainly seeking an intense, spectrally smooth source of ultraviolet light and x-radiation for use in experiments concerned with astrophysics are hopeful that synchrotron radiation will prove to be the answer. If so, they will enjoy the "bonus" that the radiation is polarized. A further bonus may exist: perhaps the experimenters will not have to measure the intensity of the synchrotron radiation (a difficult task at best), but instead may calculate it directly from the basic principles of electromagnetic theory. Heretofore there have been few light sources whose outputs could be calculated from first principles. Among these few are the blackbody, used in calibrating radiometers, the magnetic undulator, and the Čerenkov radiator, discussed in a later section.

Astronomers have been delighted to learn about synchrotron radiation since it appears to be a ready-made explanation of the high degree of polarization found in light from the Crab nebula. As indicated in Chapter 11, this nebula is now believed to be, in some sense, a vast synchrotron.

MAGNETIC UNDULATOR

In 1953, H. Motz, W. Thon, and R. N. Whitehurst decided to test the hypothesis that visible light would be given out by a horizontal stream of high-speed electrons passing through an alternating sequence of upward-directed and downward-directed magnetic fields. They reasoned that the electrons, being urged right, left, right, etc., by the successive fields, should follow an approximately sinusoidal path in a horizontal plane and should emit electromagnetic radiation polarized linearly and horizontally. To test the hypothesis, they constructed a *magnetic undulator* consisting of a straight series of oppositely directed magnets which were spaced 2 cm apart. When a slender stream of 100-MeV electrons was directed along the undulator, light of the predicted wavelength was observed. As expected, the light was linearly polarized and the electric vibration direction was horizontal.

PURCELL-SMITH GENERATOR

In 1953 E. M. Purcell and S. J. Smith produced polarized light by means of an as-yet-unnamed device that should perhaps be called a microscopic electrostatic undulator. The only important piece of equipment was an aluminum diffraction grating having 15,000 grooves per inch. To produce linearly polarized light, the experimenters directed a slender beam of 0.3-MeV electrons parallel to the plane of the grating but perpendicular to the grooves. When the beam was adjusted so as to clear the grooved surface by less than 0.1 mm, light was emitted. Why? Because as each electron traveled along just in front of the grating, it induced in the nearest portion of the grating a mirror-image charge which was forced to travel along, up hill and down dale, across the grooves so as to keep up with the electron. Each electron and its mirror-image charge constituted an electric dipole, a fluctuating dipole, thanks to the hills and dales. Such a dipole necessarily emits radiation, and the electric vector should be parallel to the axis of the dipole. This is exactly what Purcell and Smith found. Incidentally, the electron speed was so high, and the Doppler effect so large, that when the experimenters looked almost straight upstream, they saw blue light of a certain frequency; but from a vantage point far off to the side they observed light having only about two-thirds the frequency, i.e., red light. Here, then, is a light source which is colored radically differently to different observers all fixed in the same coordinate system!

ČERENKOV RADIATOR

Any transparent body in which charged particles are traveling at sufficiently high speed is a Čerenkov radiator. The transparent body may be gaseous, liquid, or solid, and any type of charged particle may be used. The speed, however, is critical: if it exceeds the speed of light in the body, linearly polarized light is emitted; if it does not exceed it, no light is emitted.

The mechanism of emission of the Čerenkov radiation is well understood. When an electron or other charged particle travels within, say, a block of glass, it distorts the charge distribution within each atom situated close to the line-of-flight, and as the electron goes on past, the distortion "relaxes." If the electron is traveling slowly, the waxing and waning of the electrical distortions fail to produce electromagnetic radiation: the pattern of the hundreds of distorted atoms is spherically symmetric, and no radiation is emitted. If the electrons are traveling faster, say at one-tenth the speed of light, the pattern continues to have cylindrical symmetry but has fore-and-aft asymmetry; yet even in this case there is no radiation, because the various Huygens wavelets that originate at various portions of the pattern produce, at any remote observation point, destructive interference. When, however, the electrons travel faster than the speed of light in the body, the wavelets reinforce one another in certain directions, and light *is* emitted. The permissible directions correspond to the straight elements of a cone such as Cone 1 shown in Fig. 12-4. The half-angle θ of the cone can be found by drawing Huygens wavelets about two points P_1 and P_2 on the electron path; the radii of the wavelets are chosen with due regard to the refractive index n of the glass, the time taken for the electron of speed v to go from P_1 to P_2, and the speed c/n of the light in the body. The construction shows that $\theta = \text{arc cos } (c/nv)$. The maximum

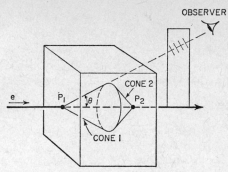

FIG. 12-4 Production of Čerenkov radiation within a transparent block by means of an electron that entered the block at P_1 and is about to leave the block at P_2. The propagation directions are parallel to the elements of Cone 1, and the wavefront constitutes Cone 2. The electric vibration direction lies in the plane that includes the observer and the path of the electron.

possible speed of the electron is c, and the maximum value of θ is arc cos $(1/n)$, which is about $48°$ for glass of index 1.5. If the speed of the electron drops below c/n, there is no radiation.

The wavefront, being everywhere perpendicular to the propagation directions, is itself conical; in Fig. 12-4 it is called Cone 2.

The electric vibration direction of the Čerenkov radiation appears different to differently situated observers. An observer situated above or below the axis of the cones finds the vibration direction to lie in a vertical plane, while an observer off to one side finds the vibration direction to lie in a horizontal plane. A rule that is always valid is that the observer finds the vibration direction to lie in the plane that includes him and the path of the electron. The reason for this is that the axis of the distortion produced in an individual atom necessarily lies in the plane that includes the atom and the electron path.

Čerenkov radiators are ideal "speedometers" for use in high-energy physics experiments. A gas-filled Čerenkov radiator can be used to distinguish dramatically between particles traveling faster than, say, 99 percent of the speed of light in vacuum and those traveling slower than this. By adjusting the gas pressure, the investigator can achieve almost any desired value of threshold of speed.

Bibliography

GENERAL

American Institute of Physics, *Polarized Light: Selected Reprints,* American Institute of Physics, New York, 1963, 103 pp.

S. S. Ballard, E. P. Slack, and E. Hausmann, *Physics Principles,* Van Nostrand, Princeton, N. J., 1954. A college textbook; see Chapter 33 especially.

R. W. Ditchburn, *Light,* Interscience (Wiley), New York, 2nd ed., 1963. See Chapters 12 and 16 especially.

F. A. Jenkins and H. E. White, *Fundamentals of Optics,* McGraw-Hill, New York, 3rd ed., 1957. Chapters 24-28.

W. A. Shurcliff, *Polarized Light: Production and Use,* Harvard University Press, Cambridge, Mass., 1962, 207 pp.

J. M. Stone, *Radiation and Optics,* McGraw-Hill, New York, 1963. See Chapters 13 and 17.

R. A. Waldron, *Waves and Oscillations,* Van Nostrand, Princeton, N. J., 1964 (Momentum Book #4), 135 pp.

C. D. West and R. C. Jones, "On the Properties of Polarization Elements as Used in Optical Instruments. I. Fundamental Considerations," *J. Opt. Soc. Am.* **41,** 975 (1951).

E. A. Wood, *Crystals and Light,* Van Nostrand, Princeton, N. J., 1964 (Momentum Book #5), 160 pp.

SPECIAL TOPICS

R. A. Beth, "Mechanical Detection and Measurement of the Angular Momentum of Light," *Phys. Rev.* **50,** 115 (1936).

G. R. Bird and M. Parrish, Jr., "The Wire Grid as a Near-Infrared Polarizer," *J. Opt. Soc. Am.* **50,** 886 (1960).

E. G. Coker and L. N. G. Filon, *A Treatise on Photo-elasticity,* Cambridge University Press, New York, 2nd ed., rev. by H. T. Jessop, 1957, 720 pp.

D. H. Frisch and A. M. Thorndike, *Elementary Particles,* Van Nostrand, Princeton, N. J., 1964 (Momentum Book #1), 153 pp.

J. V. Jelley, *Čerenkov Radiation and Its Applications,* Pergamon Press, London and New York, 1958, 304 pp.

E. H. Land, "Some Aspects of the Development of Sheet Polarizers," *J. Opt. Soc. Am.* **41,** 957 (1951).

T. H. Waterman, "Polarized Light and Animal Navigation," *Scientific American* for July 1955, Vol. 193, p. 88.

Index